The Fire
THAT SHOULD NEVER BE:
AI, Prevention and Resilience

The Fire
THAT SHOULD NEVER BE:

AI, Prevention and Resilience

Dr. Surajit Lahiri

SHROFF PUBLISHERS & DISTRIBUTORS PVT. LTD.
Navi Mumbai

The Fire That Should Never Be: AI, Prevention and Resilience

By Dr. Surajit Lahiri

First International Edition: June 2026

ISBN: 978-93-6808-094-7

TRADEMARKS

All brand names, product names, service marks, trademarks, and registered trademarks mentioned in this book, on its cover, or in its marketing materials are the property of their respective owners. The publisher and author(s) have made every effort to ensure that all such names are correctly identified and credited. The use of any trademark or service mark in this publication is for identification and descriptive purposes only and is not intended to infringe on the rights of the trademark holder.

NOTICE OF NON-AFFILIATION

This book is an independent publication written by the author(s) and published by Shroff Publishers & Distributors Pvt. Ltd. It is not affiliated with, sponsored by, authorized by, endorsed by, or connected in any way with the owners of any trademarks mentioned herein, including but not limited to Oracle Corporation, The Python Software Foundation, The Apache Software Foundation, Docker Inc., Microsoft Corporation, Google LLC, or any of their subsidiaries or affiliates. Any mention of a specific product or service is not an endorsement of that product or service. All information is provided on an "as is" basis without warranty of any kind.

Printed at Decora Book Prints Pvt. Ltd., Mumbai • Published by Shroff Publishers and Distributors Pvt. Ltd. B-103, Railway Commercial Complex, Sector 3, Sanpada (E), Navi Mumbai 400705 • Tel: (91 22) 4158 4158 WhatsApp +91 7304487700 • E-mail: spdorders@shroffpublishers.com • www.shroffpublishers.com

Dedication

This book is dedicated to the voiceless animals–the silent sufferers of disaster, climate change, and human expansion.

They do not feature in balance sheets. They do not attend emergency briefings. They do not vote, litigate, or escalate.

Yet they are often the first to sense danger and the first to pay the price when systems fail to listen.

Their vulnerability, displacement, and quiet resilience sit at the centre of this book. Not as sentiment, but as signal. Not as metaphor, but as reality.

Every landscape remembers what institutions forget. Every ecosystem carries intelligence long before alarms sound.

This book is written in recognition of that truth and in responsibility to it.

Foreword

By Dr. Joaquin Jay Gonzalez III, Vice Provost for Global Affairs and Founding Dean of Golden Gate University (GGU) Worldwide

In an age defined by accelerating climate volatility and increasingly complex technological systems, the greatest risks facing our societies rarely arrive as sudden surprises. Instead, they gather slowly through small signals, subtle shifts, and patterns that emerge quietly across landscapes, infrastructures, and institutions. By the time these patterns become undeniable, the opportunity to prevent harm has often already passed. *The Fire That Should Never Be: AI, Prevention and Resilience* invite us to confront this reality and reconsider how modern institutions perceive, interpret, and act upon early warning.

Dr. Surajit Lahiri's work arrives at a time when artificial intelligence is rapidly transforming our ability to observe the world. Sensors, satellites, lidar, radar, predictive models, and digital platforms now allow us to detect environmental stress, infrastructure vulnerabilities, and ecosystem disruptions earlier than ever before. Yet the existence of early intelligence does not automatically translate into early action. The deeper challenge lies not in technology itself, but in the institutional systems and human behaviour responsible for translating insight into credible decisions and rapid actions.

Armed with his knowledge and applications from Golden Gate University Worldwide's Doctor of Business Administration (DBA) in Emerging Technologies with specialization in AI, Dr. Lahiri addresses that challenge with clarity, reflection, and experience. Rather than presenting AI as a technological solution in search of application, Dr. Lahiri situates it within the broader architecture of governance, economics, and leadership. He shows that prevention often fails not because signals are invisible, but because institutions are structured to act only after thresholds are crossed–after ambiguity disappears and events provide undeniable justification. By that stage, prevention has already given way to response.

A central contribution of Dr. Lahiri's work is its exploration of the economics of prevention. Traditional decision frameworks struggle to value actions whose success leaves no visible trace. When a disaster does not occur, the benefits of early intervention rarely appear in conventional metrics. Lahiri challenges this limitation by reframing value in terms of resilience, volatility reduction, and strategic optionality–concepts that help institutions recognise the economic logic of acting before crisis forces alignment.

Equally important is Dr. Lahiri's recognition that prevention is ultimately a question of trust. Early signals must be trusted by decision-makers. Governance systems must legitimize action under uncertainty. Financial actors must believe that institutions can manage emerging risks responsibly. Without this trust, intelligence accumulates without authority, and warnings remain trapped within systems designed for a different era.

At its core, *The Fire That Should Never Be* is a book about leadership in a time when the future has become more visible yet more uncertain. It challenges executives, policymakers, regulators, and practitioners to reconsider the assumptions that shape their decisions. The question is no longer whether risks can be detected early–increasingly, they can. The real question is whether institutions are prepared to act on that knowledge before disaster becomes the only justification left.

The insights offered by Dr. Lahiri extends beyond wildfire management and climate and disaster resilience. They speak to the broader challenge of governing complex systems in an era of ecological change and technological capability. As Dr. Lahiri reminds us, resilience is not built in the moment of crisis, but in the quieter period before it, when signals appear and choices still exist.

Biography

Joaquin Jay Gonzalez III, PhD., is Vice Provost for Global Affairs and Founding Dean of Golden Gate University (GGU) Worldwide. He is also Mayor George Christopher Professor of Public Administration at GGU's Edward S. Ageno School of Business. Dr. Gonzalez is the co-author/editor of more than 20 books in the intersections of emerging technologies, governance, and public policy, including *Cybersecurity, The World Remade by Artificial Intelligence, The AI Factory, and The Rise of AI Agents*. He was worked as governance for the World Bank, the Inter-American Development Bank, UN International Organization for Migration, the International City/County Management Association, and USAID. A teaching, service, and research excellence awards recipient, Dr. Gonzalez has taught at the University of San Francisco, the National University of Singapore, and De La Salle University, Manila.

Prof Jay helped establish the WASC-accredited MS in Artificial Intelligence (MSAI) at the GGU Worldwide. MSAI is offered in both English and Chinese. He helped design and implement GGU World wide's Responsible Use of AI Policy.

About the Editor

Dr. Sumitra Padmanabhan's academic and research career spans 25+ years across academia and industry. Beginning as a process control and instrumentation engineer, her journey has evolved into specializing in the application of Artificial Intelligence (AI) and Machine Learning (ML) to real-world challenges. Her focus goes beyond building intelligent solutions; her work emphasizes making AI understandable, practical, and relevant for non-technical and business audiences.

Dr. Padmanabhan did her PhD in space weather, focusing on improving the positional accuracy of GPS signals under ionospheric and geomagnetic variability-work at the intersection of satellite communications, communication physics, and machine learning. This research shaped her shift from "electrons to bits," and strengthened her commitment to data-driven approaches for solving complex, real-world problems.

Her teaching experience covers a wide range of subjects, from Electronics and core engineering to AI and Business Analytics, across undergraduate, graduate and doctoral programs. Her leadership contributions include curriculum development, academic governance, and program delivery, with an emphasis on outcome-based, industry-aligned education designed for working professionals and future leaders.

At present, she works as an Associate Director at UpGrad and an Adjunct Professor at Golden Gate University. She taught DBA–ET courses 805, 808, and 860, and course 800 in the DBA flagship program and mentored doctoral learners as a dissertation committee Chair (5 students) and Co-Chair (2 students).

With a commitment to democratizing AI education and expanding access to quality learning, I collaborated with IIM Bangalore to deliver an 8-week course on *Machine Learning for All* on the Government of India's SWAYAM platform. The course has seen more than 50K enrolments till date.

Beyond institutional roles, she mentors girl students pursuing technical education through a not-for-profit organization, supporting both their career development and key life decisions. At the heart of her work is a simple belief: *"Share what you know, and give back where you can, always, in whatever way possible"*.

Strategic Advisor

Dr. Rajendra B. Patil is a globally recognized academic leader, AI researcher, entrepreneur, and strategic innovator with over two decades of experience at the intersection of education, technology, and institutional transformation. He serves as Director at Uma Krishna Shetty Institute of Management Studies, Mumbai, Director–Institutional Growth & Strategies at Anna Leela College, Mumbai, and Visiting Faculty at Jamnalal Bajaj Institute of Management Studies (JBIMS), Mumbai–one of India's premier management institutions.

A key thought leader in the post-NEP 2020 transformation of Indian education, Dr. Patil has led large-scale initiatives in academic restructuring, accreditation excellence, and governance modernization. His work focuses on building future-ready, multidisciplinary institutions that integrate emerging technologies, industry collaboration, and outcome-driven learning models.

Dr. Patil has pioneered the design and implementation of next-generation academic programs in Artificial Intelligence, Machine Learning, Data Science, Clinical Data Science, and Cyber Security–positioning institutions to meet global workforce demands. His contributions extend to architecting AI-driven academic ecosystems, enabling digital transformation through automation, online assessments, and intelligent learning platforms.

An accomplished scholar, Dr. Patil has published over 40 peer-reviewed research papers, authored six academic books, and holds multiple patents spanning Artificial Intelligence, Healthcare Technologies, Cloud Computing, and Intelligent Automation. As a recognized Ph.D. supervisor at the University of Mumbai, he mentors doctoral researchers contributing to cutting-edge advancements in AI and data sciences.

On the global stage, Dr. Patil has delivered more than 1,000 executive training programs and workshops for leading organizations, including NTUC Singapore, Samsung, and the Aditya Birla Group. His expertise spans Generative AI, advanced analytics, business intelligence, and enterprise digital transformation, making him a sought after advisor for institutions, corporates, and innovation ecosystems.

Blending academic rigor with entrepreneurial vision, industry relevance, and strategic foresight, Dr. Rajendra B. Patil continues to influence the evolution of AI-enabled education and digital transformation worldwide, shaping not just institutions, but the future of learning itself.

Acknowledgement

This work may carry a single name on its cover, but it was never created alone.

Long before this book took form on paper, it was already being shaped quietly by the people who stood beside me–through encouragement, patience, belief, and sacrifice. While my name appears on the cover, this work carries within it the presence of many lives that made its creation possible.

My deepest gratitude goes to Dr. Dakshinamurthy Kolluru, my motivation and my silent mentor. His guidance was never loud, yet it was constant. Through quiet wisdom, personal example, and unwavering faith in my purpose, he helped shape the direction of this journey in ways that words cannot fully capture.

I am sincerely thankful to Dr. Sumitra Padmanabhan for her invaluable editorial contribution. Her thoughtful review, attention to detail, and ability to refine the narrative have played a significant role in shaping this manuscript into its present form.

I am especially grateful to Dr. Seema Purohit, whose mentoring helped weave this work into a coherent whole. Her guidance went beyond structure and form–she helped me see how the technical, the human, and the ethical threads of this journey could come together as a single narrative. Her ability to bring clarity and continuity to complex ideas played a defining role in shaping the voice and flow of this book.

I would also like to express my sincere gratitude to Dr. Joaquin "Jay" Gonzalez III, for his thoughtful guidance and academic insight throughout this journey. His perspective helped strengthen the intellectual foundation of this work, and his encouragement played an important role in shaping its direction and purpose.

I would like to extend my sincere appreciation to Dr. Rajendra Patil for his role as a strategic advisor. His perspective, guidance, and ability to connect ideas to broader strategic and real-world contexts added significant depth to this work and helped shape its practical relevance.

To my family, this journey would not have been possible without your patience, strength and unwavering belief in me. You are the constant strength behind every effort I make. The sacrifices you made live within every page of this book.

I bow with reverence to the blessings of my parents, who now watch over me from the heavens. Their values remain woven into every decision I take and every purpose I pursue.

This book is, in many ways, a reflection of all of you.

Abstract

In the early hours of many Australian mornings, long before smoke is visible or sirens begin to sound, a fire is already forming somewhere in the bush. A fallen powerline, a sudden gust of wind, dry grass under extreme heat–and within minutes, a small spark becomes an unstoppable force. Floods follow a similar pattern. They no longer arrive slowly or predictably. They arrive suddenly, rise fast, and leave lasting damage in their wake. Nature has changed its rhythm. Our systems for managing disasters, however, are still struggling to keep up.

The Fire That Should Never Be: AI, Prevention and Resilience is written from this reality. It is not a book about technology alone. It is a book about anticipation, responsibility, protection, and balance–between human progress and the natural world. It explores how Artificial Intelligence, when designed and used with care, can help us detect danger earlier, act faster, and protect both lives and landscapes more effectively than ever before.

Set primarily in the Australian context but relevant globally, the book brings together satellite imagery, environmental sensors, utility infrastructure data, climate models, and aerial monitoring into one connected intelligence ecosystem. It shows how early signals–rising temperatures, shifting winds, unusual animal movement, line faults, soil saturation can be turned into meaningful warnings before devastation begins. Throughout the book, strong emphasis is placed on transparency, explainability, and ethical responsibility, ensuring that AI systems remain understandable, trustworthy, and accountable.

What makes this work different is its strong focus on practical value and financial reality. Disaster prevention is too often seen as a necessary expense rather than as a strategic investment. This book challenges that thinking. It introduces a clear, structured way to measure the economic value of prevention–from avoided infrastructure damage and reduced emergency response costs to insurance benefits, operational efficiencies, and the long-term value of protected ecosystems and wildlife.

The research behind the book draws on technical system design, real data from global environmental platforms, and in-depth conversations with professionals from utilities, emergency services, AI companies, finance, sustainability, and academia. Their collective insights shape a practical roadmap for moving from reactive disaster response to proactive, intelligence-driven prevention.

At its heart, this book tells a simple but powerful story: disasters do not begin with flames or floodwaters–they begin with signals. If we learn to listen to those signals through intelligent systems, we can act earlier, reduce harm, protect wildlife, secure infrastructure, and create safer communities. This book offers a grounded, practical vision of how AI can serve not just industries and governments, but the planet itself–responsibly, ethically, and with lasting impact.

TABLE OF CONTENTS

Chapter 1
The Moment We Missed

Why Disasters Are No Longer Sudden—And Institutions Still Are?

The system saw it coming.

It simply did not act.

Long before the first flame appeared, the signals were already there–scattered across data streams, embedded in patterns, visible to those trained to look. Nothing was hidden. Nothing was missed.

What was missing was permission.

The problem is not detection.

It is permission.

And that is where this story begins.

Harnessing AI for impact does not begin with algorithms, dashboards, or alerts. It begins with institutions learning how to act early without being punished for it later. This book is not a manual on firefighting tactics or wildlife rescue operations. It is an examination of the systems that decide whether those interventions ever occur early enough to matter. Artificial intelligence already detects fire risk, environmental stress, and infrastructure failure far earlier than human institutions can respond. The persistent failure has not been technological. It has been institutional–rooted in how risk is interpreted, how authority is granted, and how capital is allowed to move before crisis makes action unavoidable.

The profitable prevention of fire and the protection of wildlife depend not on better prediction alone, but on whether early intelligence can be translated into legitimate, fundable decisions. AI creates impact only when its signals are trusted, governed, and sustained economically–especially when success produces no visible event to justify the spend. Throughout this book, fire prevention and wildlife protection are treated not as isolated outcomes, but as downstream consequences of institutional design choices. When anticipatory intelligence is granted legitimacy early, capital can be structured to wait, optionality can be preserved, and ecosystems–human and natural–can remain intact without ever becoming headlines. This is what it means to harness AI for impact: not to react faster to disaster, but to make prevention economically and institutionally viable before disaster is required to prove its value.

No one remembers the moment when things first began to feel wrong.

There is no timestamp for it. No alert that failed to trigger. No meeting where someone said, this is the point where we should have acted. In hindsight, people will search for that moment. They will argue about it. They will draw lines on charts and mark inflection points that look obvious only after the fact.

But that is not how it happened.

What people remember instead is a stretch of time when nothing quite resolved the way it used to. Conditions lingered longer than expected. Small anomalies stopped disappearing.

Systems held, but only just. Everything remained operational, compliant, and explainable–and yet, a quiet unease began to spread across teams that were used to certainty.

This is how modern disasters begin.

Not with failure, but with endurance.

Figure 1.1 The fire begins long before the first flame-in data, patterns, and ignored signals.

Inside an operations centre, a team monitors the same screens they have always monitored. The data looks familiar. The metrics remain within limits. Alerts trigger and clear as designed. If someone walked in for the first time, then nothing would appear alarming.

But the people who have been here for years sense a shift.

It is not dramatic enough to escalate. Not clear enough to justify intervention. It is simply persistent. The same patterns reappear across days, then weeks. The environment behaves slightly differently. Infrastructure shows signs of fatigue earlier than expected. Human workarounds quietly increase to maintain stability.

No rule is broken. No protocol violated.

The system is doing exactly what it was designed to do. That is the problem.

At a management level several layers removed from the screens, decisions are being made with equal professionalism. Reports are reviewed. Risk registers updated. Mitigation plans refined. Everything is defensible. Every action can be explained later if needed.

This matters deeply.

Modern institutions are built not just to function, but to justify. Decisions must withstand scrutiny from regulators, boards, auditors, insurers, and courts. Acting too early can be as dangerous, reputationally and financially, as acting too late.

So, caution becomes policy. And policy becomes habit.

In a boardroom, the conversation sounds rational.

There is no denial here. Leaders acknowledge increasing volatility. They discuss climate variability, infrastructure exposure, supply chain fragility. They approve studies. They fund pilots. They ask for more data.

What they do not do is act decisively–because nothing yet demands it.

There is no event.

The Meeting Where Nothing Happened

A senior manager sat through the meeting.

The slides were clean. The metrics were within range.
Everything looked stable–exactly as expected.

Someone mentioned "increasing variability," almost in passing. Another suggested "keeping an eye on it."

Heads nodded.

No one disagreed.
No one escalated.

The decision was not to ignore the risk.

The decision was to wait.

And because it was reasonable, documented, and collectively agreed, it became invisible.

Weeks later, no one would remember this meeting as a decision point.

But it was.

There is only accumulation.

This is the first misalignment that defines our era:

- risk no longer arrives as an event, but institutions still wait for one.

For decades, this waiting made sense. Disasters were sudden. Signals were noisy. Acting too early created unnecessary cost and political backlash. Thresholds existed for a reason.

But the nature of risk has changed.

Environmental systems no longer reset cleanly. Infrastructure ages under compound stress. Human adaptation masks degradation until margins vanish. Financial exposure builds silently through insurance repricing, capital hesitation, and regulatory tightening.

None of these trips an alarm.

All of it matters.

In the field, experienced teams adapt instinctively. They change how they work. They slow down. They double-check. They share informal knowledge about what is different now. Their competence is impressive.

It is also dangerous.

Because adaptation delays visibility.

The more capable the humans, the longer the system can operate beyond its design assumptions without appearing to fail. Stability becomes indistinguishable from strain.

From the outside, everything still looks fine.

This is where the real failure occurs–not in data, not in intent, not in competence, but in timing.

By the time conditions become undeniable, the range of options has already narrowed. Decisions are still possible, but they are no longer free. Capital must move quickly instead of deliberately. Governance shifts from anticipation to defence. Human judgement is replaced by urgency.

Institutions call this response.

What it really is, is catch-up.

The tragedy is that the signals were there.

Not as predictions. Not as certainties. But as patterns–weak, persistent, interacting patterns that quietly described a direction of travel.

The problem was never visibility.

It was legitimacy.

Case Study

The Fire That Was Already There: Camp Fire (California, 2018)

The signals were already there.

Dry vegetation. Strong seasonal winds. Aging transmission lines running through high-risk terrain. None of this was unknown. It had all been seen before–individually, repeatedly.

On November 8, 2018, a fault in a power line triggered what became the deadliest wildfire in California's history. The town of Paradise was destroyed within hours.

Nothing about the conditions was surprising.

What was missing was early action.

Shutting down power would have been disruptive, costly, and difficult to justify if nothing happened. So the system waited.

The fire did not begin with ignition. It began with a decision not to act.

Executive Takeaway:

The greatest risk is not what we fail to see–it is what we choose not to act on.

The Silence Between Signal and Authority

Why Early Knowledge Rarely Becomes Permission?

Across modern institutions, legitimacy is granted retrospectively. Action is justified after consequence. This works when risk is episodic and bounded. It fails when risk accumulates gradually across systems that no single authority owns.

No one is authorised to act on "almost".

No one is rewarded for preventing something that never happens.

Acting early risks embarrassment if nothing happens. Acting late risks harm but harm distributes blame more evenly.

What makes this moment different from the past is not the scale of risk, but its shape.

Risk no longer spikes.

It settles.

It lingers.

It converges.

And convergence changes everything.

This book begins here-not with technology, and not with catastrophe, but with the quiet moment when institutions realise that their strengths have become constraints.

The systems designed to protect them from rash action now prevent timely action. The processes built to ensure accountability now delay responsibility.

This is not a story about failure.

It is a story about design lag.

The chapters that follow will show how this lag emerged, why it persisted even in competent organisations, and how a different operating logic-one that integrates early intelligence, legitimate governance, and sustainable capital can close the gap.

But first, we must understand what it feels like to miss the moment.

Because no institution ever decides to fail.

It simply waits for a signal that arrives too late to be useful.

Before the Fire

The Signal That Arrived Early-And Went Unheard

Nothing happened that day.

Somewhere inside the system, someone notices the moment anyway.

They are not looking for danger. They are looking for continuity. The screens in front of them show stability. Values remain within tolerance. Alerts clear themselves. Everything that should be happening, is happening.

And yet, something feels misaligned.

It is not fear. It is not urgency. It is the quiet sense that a pattern has shifted just enough to deserve memory. They mark it privately–not in a report, not in a system–just enough to remember that this was the first time it did not resolve the way it usually does.

There is nothing to escalate. The system does not recognise unease as intelligence.

There were no sirens, no emergency declarations, no disruption to daily routines. Life continued as normal.

From the outside, the system appeared stable.

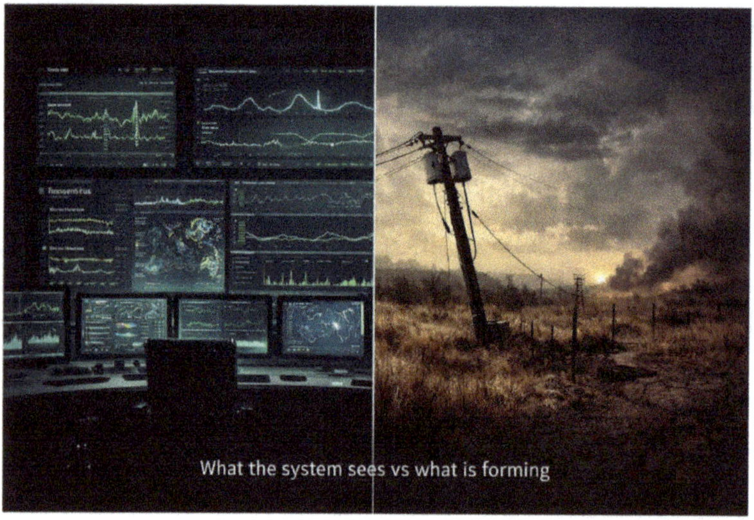

Figure 1.2 Stability is often an illusion–risk is already forming

But stability, in high-risk environments, is often an illusion created by timing.

Long before damage becomes visible, conditions quietly assemble. Heat accumulates across the landscape. Wind behaviour grows erratic. Vegetation moisture drops below seasonal norms. In a utility corridor cutting through dry terrain, a minor fault registers briefly before clearing itself.

No threshold is crossed.

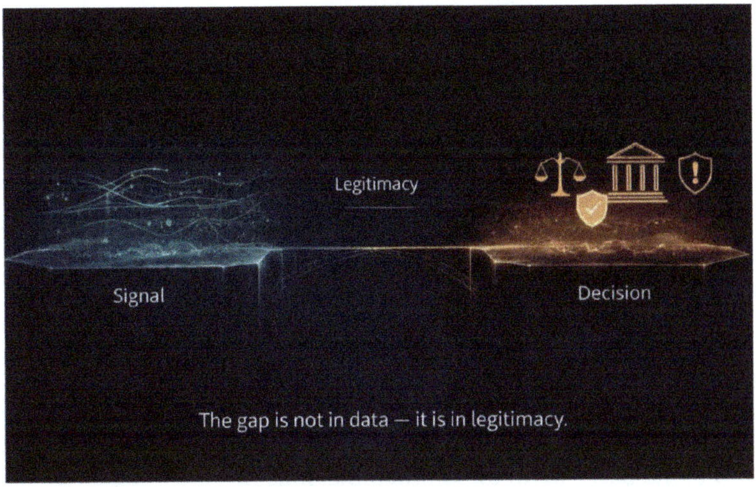

Figure 1.3 Signals flow. Decisions wait. Legitimacy is the missing bridge

No escalation protocol is triggered.

No decision is required.

In operational terms, this is success.

And yet, this is the moment that matters most–the final window where prevention remains possible, and intervention would still be inconvenient rather than catastrophic.

This book begins here–not with fire, but with absence. With the quiet period before consequence arrives, because this is the phase institutions are least prepared to treat as meaningful.

Disasters do not begin when response systems activate. They begin earlier, in spaces where signals are present but authority is not.

The idea that disasters are sudden is comforting. It suggests inevitability. It absolves systems of responsibility for anticipation and places the burden of action firmly on response. If nothing could have been done earlier, then nothing was missed.

But this idea no longer holds.

Wildfires, floods, and infrastructure failures increasingly follow recognisable trajectories. They emerge from patterns–environmental, mechanical, and operational that develop over time. These patterns are detectable. In many cases, they are already detected.

What fails is not sensing.

What fails is translation–the movement from signal to decision.

In the days before a fire, data flows continuously. Environmental sensors record temperature, humidity, and wind behaviour. Satellite imagery captures heat signatures and vegetation stress. Infrastructure systems monitor load fluctuations, faults, and degradation. Predictive models update risk surfaces based on changing conditions.

None of this is hypothetical. It is routine.

Yet despite this intelligence, early action remains rare.

The reason is not ignorance. It is institutional design.

Early signals live in an uncomfortable space.

They are neither benign nor decisive. They suggest increasing probability rather than certainty. Acting on them requires judgment before confirmation and judgment without institutional protection is risky.

In most organisations, authority is structured to activate after thresholds are crossed. Escalation protocols, emergency declarations, and response funding are triggered by events that are already undeniable. This architecture makes sense in environments where false alarms are costly and public trust is fragile.

But in dynamic risk contexts, this design introduces delay precisely when time matters most.

Early signals are therefore treated cautiously. They are logged, reviewed, monitored, and discussed but rarely acted upon. They remain informational rather than operational.

The moment to raise it comes and goes.

They imagine the questions that would follow–what exactly has changed, what action is being proposed, what evidence supports the concern. None of these questions are unreasonable. All of them require certainty that does not yet exist.

The note is rewritten.

Then softened. Then removed entirely.

The Email That Was Never Sent

An operations engineer noticed it before anyone else did.
Not because he was looking harder, but because he had seen this pattern before.

The temperature variance was small. The wind shift was explainable. The fault cleared itself.

Everything, individually, made sense.

Together, it did not.

He opened a draft email.

"Not urgent," he wrote first. Then deleted it.
"Possibly worth monitoring," he tried next. Then softened it again.

He paused.

What exactly was he asking people to act on?
There was no threshold crossed. No rule broken. No model breach.

Only a signal–structured, informed, but not yet defensible.

The email was saved. Then closed.

Nothing in the system recorded that moment.

But weeks later, when the fire did come,
he would remember it with uncomfortable clarity.

Nothing in the system records this decision.

Competence remains intact. Prevention quietly slips out of reach.

By the time a signal becomes authoritative, the system has already shifted from prevention to response.

Practitioners recognise this pattern instinctively.

Control rooms are filled with dashboards that glow continuously. Alerts arrive at all hours. Operators learn to distinguish between noise and urgency, between anomalies that can be ignored and those that demand immediate attention. This filtering is essential. Without it, systems would grind to a halt.

The unintended consequence is that early warnings often look indistinguishable from background variation.

When systems are calibrated to respond to threshold breaches rather than trajectory changes, they become highly effective at managing emergencies and poorly equipped to prevent them.

This is not negligence. It is optimisation.

Institutions optimise for stability. They prioritise continuity, credibility, and defensibility. A system that escalates too often risks being ignored when escalation truly matters. A leader who acts too early risks being judged alarmist.

And so, caution becomes the default posture.

Caution, however, is not neutral.

Waiting for confirmation trades early, low-impact intervention for late, high-impact response. This trade is rarely explicit. It is embedded quietly in procedures, escalation chains, and cultural norms.

Early intervention requires explanation. It requires leaders to justify decisions that may never appear necessary in hindsight. It concentrates accountability at the moment of action.

Waiting, by contrast, diffuses responsibility. If damage occurs, then response can be justified. If nothing happens, then no explanation is required.

This asymmetry shapes behaviour.

Systems rarely ask, "What happens if we act early?"

They ask, "What happens if we act too soon?"

The answer to the second question is immediate and personal. The answer to the first is abstract and collective.

As a result, early action is deferred not because it is irrational, but because institutions are not designed to reward it.

For practitioners, this tension is lived rather than theoretical.

Shutting down infrastructure before visible threat disrupts customers and revenue. Issuing warnings too early risks public scepticism. Mobilising resources in advance invites scrutiny from oversight bodies and media commentators who arrive after the fact, armed with hindsight.

These consequences are tangible and attributable.

The benefits of prevention, however, are counterfactual. They exist only as stories about what did not happen. They are difficult to quantify and easy to dismiss.

In post-event reviews, successful prevention often disappears from the narrative. What remains is a response to the visible effort, the documented expenditure, and the measurable activity.

Over time, this imbalance reinforces reactive behaviour, even as risks accelerate. After every disaster, reports are written.

They catalogue timelines, response effectiveness, coordination challenges, and recovery outcomes. They analyse what went wrong once the event was underway.

Rarely do they linger on the days or hours before the period when conditions were forming, signals were present, and options still existed.

Those moments are harder to document. They involve judgement calls that were never made, thresholds that were never crossed, and decisions that were never required.

And yet, those moments determine resilience.

They are the moments this book insists on holding in view.

At some point in recent history, environmental risk crossed a quiet threshold.

Wildfires ceased to be rare, seasonal disruptions and became persistent, compounding threats. Floods followed similar trajectories. Infrastructure systems designed for historical patterns began to fail under new conditions.

What did not change as quickly were the assumptions underpinning institutional decision-making.

Risk continued to be treated as episodic. Authority remained designed to activate after confirmation. Response continued to be prioritised over anticipation.

The result is a widening gap between the speed of risk and the speed of decision.

This gap is not a failure of intelligence. It is a failure of design.

This section has deliberately stayed with the moment before damage–the phase where nothing appears to be happening, and therefore nothing feels urgent.

Because this is where resilience is either built or lost.

Before the fire, there is still choice.

Before the fire, action is still possible.

Before the fire, the cost of intervention is lowest–and the cost of hesitation is least visible.

It is also the moment institutions are least prepared to recognise as decisive.

If early danger were loud, then institutions would not struggle to respond.

The difficulty lies in the fact that risk whispers long before it shouts.

Inside operational environments, signals rarely arrive as clear warnings. They appear as marginal deviations, transient anomalies, or patterns that only make sense when viewed over time. A sensor reading that is slightly higher than usual. A fault that clears itself. A forecast that nudges probability upward without crossing any defined threshold.

Individually, these signals are unremarkable. Collectively, they are meaningful.

Yet collective meaning is precisely what most systems are least equipped to interpret in real time.

In practice, early signals enter organisations through narrow channels. They are seen first by specialists–operators, analysts, and engineers whose roles are defined by monitoring rather than decision-making. Their responsibility is to observe, not to act. To report, not to escalate.

This separation is deliberate. It preserves control and prevents unnecessary disruption. But it also creates distance between intelligence and authority.

As signals move upward, they lose urgency.

By the time they reach those empowered to act, they are often abstracted, summarised, or contextualised away. What began as a pattern becomes a line item. What felt concerning at the edge becomes manageable at the centre.

This is not failure. It is translation and translation always involves loss.

Most institutions rely on thresholds to decide when something matters. Thresholds are comforting because they offer clarity. Cross the line, and action is justified. Stay below it, and caution is defensible.

The problem is that many modern risks do not respect static thresholds.

Wildfire risk, for example, does not emerge from a single variable crossing a predefined limit. It emerges from convergence–heat, wind, dryness, infrastructure stress, and human presence interacting in ways that amplify one another. Each variable may remain within acceptable bounds, even as the system as a whole becomes fragile.

Threshold-based systems struggle in such environments because they are designed to respond to events, not to trajectories.

Early signals, by definition, sit below thresholds. They indicate direction rather than arrival. Acting on them requires judgement and judgement is harder to defend than compliance with predefined rules.

As a result, systems wait.

Waiting is not always conscious.

Often it takes the form of "continued monitoring." A decision to revisit the issue later. A request for additional data. A preference to see whether the signal persists.

These responses feel reasonable. They buy time. They reduce the risk of acting unnecessarily.

What they also do is narrow the window for prevention.

In dynamic environments, time does not stand still while institutions deliberate. Conditions continue to evolve. What was preventable becomes manageable. What was manageable becomes irreversible.

By the time certainty arrives, options have already collapsed.

There is a deeper reason early signals struggle to prompt action: they challenge narratives of control.

Modern organisations are built around the assumption that risk can be managed through planning, forecasting, and response. Early warnings disrupt this assumption by revealing uncertainty that cannot yet be resolved.

They force leaders to confront questions that do not have clean answers:

How confident is confident enough?

Who bears responsibility, if this proves unnecessary?

How will this decision look, if nothing happens?

These are not technical questions. They are institutional ones.

And institutions, like individuals, prefer clarity to ambiguity.

For practitioners, the emotional experience of early risk is often one of unease rather than alarm.

There is a sense that something is "off," without being able to point to a single cause. This discomfort is difficult to communicate upward. It lacks the precision that formal escalation demands.

As a result, it is often softened in translation.

What feels concerning becomes "worth watching." What feels risky becomes "within tolerance." What feels urgent becomes "noted."

Language matters here. The words used to describe early risk shape how it is perceived. Neutral phrasing calms systems. Cautious language reassures decision-makers. Over time, this linguistic smoothing dulls the edge of warning.

By the time language sharpens, the system is already under stress.

Another challenge lies in fragmentation.

Early signals rarely belong to a single domain. Environmental data sits with one agency. Infrastructure telemetry with another. Weather forecasting with a third. Emergency services monitor impacts rather than precursors.

Each sees part of the picture. Few see the whole.

Without integration, no single signal appears decisive. Responsibility for action remains unclear. Decisions are deferred not out of neglect, but out of jurisdictional ambiguity.

This fragmentation is often reinforced by governance structures that reward compliance within silos rather than collaboration across them.

In such systems, early risk becomes everyone's concern–and therefore no one's responsibility.

The irony is that many of the tools needed to act earlier already exist.

Data is abundant. Models are increasingly capable. Patterns are visible to those trained to see them.

What is missing is not intelligence, but permission.

Permission to act before certainty. Permission to prioritise trajectory over threshold. Permission to treat early discomfort as meaningful rather than inconvenient.

Without this permission, early signals remain trapped–detected, acknowledged, and ultimately absorbed into normality.

This absorption is dangerous because it normalises escalation.

As conditions worsen gradually, each step feels tolerable relative to the last. Systems adapt to rising risk by redefining what counts as acceptable. What would have triggered concern months earlier becomes the new baseline.

By the time the situation demands action, it does so abruptly–not because risk arrived suddenly, but because tolerance eroded quietly.

This phenomenon explains why disasters often feel shocking even when they were predictable. The system adjusted continuously, until it could no longer do so.

Before the fire, there is still space to choose.

But that space exists only briefly, and it exists precisely where systems are least comfortable operating–in ambiguity, without certainty, and under imperfect information.

This is the paradox at the heart of early risk management.

Institutions demand certainty to act, yet certainty arrives only after the opportunity to prevent has passed.

Until this paradox is addressed, early signals will continue to arrive and continue to go unheard.

Before the Fire does not argue that every early signal should trigger intervention. It argues something more measured and more demanding: that systems must be designed to recognise when early action is justified, even when it feels uncomfortable.

That design challenge sits at the centre of this book.

Built to React, Not to Anticipate

Why Modern Systems Lag Behind Modern Risk

The day does not announce itself as important. It never does. There is no marker that separates this day from the ones before it or the ones that will follow. It enters quietly, carrying with it the same routines, the same systems, the same expectations of stability. People arrive, screens light up, information begins to move. Everything works.

This is what competence looks like when nothing is visibly wrong.

Institutions trust these moments deeply. They are the moments that justify structure, investment, and hierarchy. When systems function smoothly, it affirms the belief that they are sufficient. That belief is not arrogance; it is earned through repetition. Years of normal operation build confidence not only in technology, but in judgment.

Yet risk does not need malfunction to grow. It thrives most comfortably in periods of apparent calm.

Early signals appear without urgency. They are not warnings in the way training manuals describe warnings. They do not arrive with volume or clarity. They slip in through margins–minor inconsistencies, slightly altered patterns, behaviours that feel different without being wrong. A value remains within tolerance but behaves strangely. A trend that once resolved now lingers. A condition that usually peaks now plateaus.

These signals are noticed. The system is not blind. But they do not carry authority.

Authority belongs to thresholds, and thresholds remain intact.

Thresholds are reassuring because they reduce complexity to certainty. They tell people when action is justified and when it is not. As long as the numbers stay on the correct side of the line, restraint feels not just reasonable, but responsible. Acting before that line is crossed invites explanation, scrutiny, and risk of embarrassment.

Modern institutions are shaped as much by the avoidance of embarrassment as by the avoidance of disaster.

In one control room, an operator with decades of experience pauses longer than usual. He has seen these systems through seasons of stress and quiet. He recognises patterns not as charts, but as feelings–something in the alignment feels unfamiliar. Not wrong. Not dangerous. Just different.

He scrolls back through historical data. There are similarities to periods that later became significant, but not enough to make a case. He considers raising a note. He imagines the questions it would generate. What exactly is the issue? What action would you recommend? What evidence supports this?

He cannot answer cleanly.

He deletes the note and continues monitoring.

Nothing in the system records this moment.

Elsewhere, in a field operations context, a supervisor notices how conditions on the ground feel harder than expected. Tasks take longer. Equipment behaves differently. The explanation is easy–weather variability, seasonal effects, minor logistical friction. These explanations are valid.

Still, the supervisor adjusts schedules quietly, building in more buffer than usual. No report captures why. It is framed as prudence. The adjustment is invisible to the broader system.

In a planning function, a risk register is reviewed. A rating is considered. The likelihood feels higher than last cycle, but not enough to justify a formal change. Formal changes trigger review chains, governance discussions, and scrutiny. The planner decides to wait for another cycle.

Each of these moments is small. Each is reasonable. None feels decisive.

Together, they form a pattern the system cannot see because it lives between silos.

Waiting becomes the default behaviour not because people are passive, but because waiting aligns with how authority is distributed. Authority flows downward once certainty exists; it rarely flows upward based on intuition.

Language adapts to this reality. Early descriptions are careful, balanced, professional. Words soften. "Concern" becomes "monitoring." "Risk" becomes "context." These shifts are not strategic; they are linguistic survival.

Language that does not provoke action also does not provoke resistance.

Over time, repeated exposure to uncertainty without consequence reshapes perception. What once felt marginally worrying now feels normal. Familiarity dulls urgency. People stop expecting resolution. They adapt to ambiguity as a permanent state.

This adaptation is stabilising. It allows work to continue. It also erodes the system's sensitivity to convergence.

A report circulates that is technically excellent. It explains complexity clearly. It highlights dependencies and caveats. It concludes that continued observation is appropriate. No one objects. The report does what it is meant to do–it reassures by explaining.

Explanation becomes a substitute for action.

Another subtle surprise arrives when two data sources that normally align do not. The mismatch is small. It does not invalidate conclusions. But it resists easy reconciliation. Several people notice it independently. None escalate it. There is no obvious owner for the discrepancy.

Between-functions is where responsibility goes to rest.

Preparedness begins to take centre stage. Response plans are reviewed. Exercises are run. Capabilities are reaffirmed. Preparedness feels productive. It is visible. It can be demonstrated. It signals seriousness without forcing premature decisions.

What preparedness quietly assumes is inevitability.

The organisation does not say this out loud, but behaviour reveals it. Energy flows toward responding well rather than preventing early. Readiness becomes the moral high ground.

Sustained uncertainty produces a different kind of fatigue–not exhaustion, but erosion. People grow tired of holding concerns that never crystallise. Vigilance becomes procedural. Curiosity fades. Adaptation hardens into acceptance.

Then comes the moment when the system comes closest to changing course.

In a senior discussion, someone proposes a small, reversible adjustment. Nothing dramatic. Just a slight shift in posture. For a brief moment, the room is quiet in a different way. People imagine what it would mean to act now rather than later.

Questions emerge, thoughtful and sincere. What if conditions stabilise? How would this be justified? What precedent would it set?

Each question slows momentum.

No one opposes the idea. No one champions it.

The proposal dissolves.

Relief follows. Tension dissipates. Normal rhythm returns. The system internalises the lesson without naming it: intervention was considered, and it was not necessary.

Weeks later, in a board-level discussion, a summary is presented. It is calm, measured, reassuring. Early signals are acknowledged, framed as managed. The board is satisfied. The meeting moves on.

Still later, when undeniable change finally arrives, it feels sudden. Authority consolidates instantly. Action is swift. The organisation performs exceptionally. The response is competent, disciplined, effective.

This is the moment institutions are built for.

Afterward, reviews are conducted. Data is reconstructed. Early indicators are rediscovered. The almost-decision resurfaces. It looks obvious now. People ask whether things could have been done differently.

The honest answer is complex.

Yes–because intervention was technically possible.

No–because intervention was not institutionally authorised.

The system did not fail to see. It failed to permit itself to act.

The Cost of Waiting That Never Appears

How Prevention Disappears From Institutional Memory

This is why the pattern repeats. Not because institutions are careless, but because they are structurally conservative in the wrong direction. They are built to respond to certainty, not to act under ambiguity.

Until that changes, the most important decisions will continue to occur quietly, almost made, almost remembered, and ultimately overtaken by events.

The day before disaster will always look ordinary. The people will always be competent.

The decisions will always feel reasonable at the time.

And prevention will continue to live in the narrow, uncomfortable space that institutions are least equipped to occupy.

And yet, even after the response has concluded and normal operations begin to reassert themselves, something subtle remains unsettled.

The organisation moves forward, but it does so with a faint sense that the story it tells itself is incomplete. This incompleteness is not expressed openly. It appears instead as a quiet dissonance between how people describe what happened and how it felt while it was unfolding. The official narrative is coherent, defensible, and reassuring. The lived experience was more ambiguous.

Ambiguity is difficult to archive.

As systems reset and routines resume, people begin to notice how quickly the earlier period is being compressed in memory. Days of observation, hesitation, and almost-decisions collapse into a brief prelude. What remains vivid is the moment of action–the point at which clarity arrived and authority followed.

This compression is not intentional. It is a feature of how institutions remember. Memory favours moments of intensity over stretches of restraint. It elevates action and blurs waiting.

In doing so, it quietly erases the only period where prevention was still possible.

Inside the organisation, learning begins to take shape. Lessons are discussed, documented, and distributed. They focus on improving response speed, strengthening coordination, refining thresholds, and enhancing communication. These are valuable lessons. They make the system more capable.

They do not address the earlier silence.

No one writes a lesson titled *What We Nearly Did.*

Near-decisions do not fit easily into formal learning frameworks. They are too speculative, too dependent on hindsight. Discussing them risks reopening questions that feel uncomfortable and unresolved. It suggests that different choices were possible, which in turn suggests that responsibility could have been exercised differently.

Responsibility is easier to assign in moments of action than in moments of hesitation.

As time passes, the almost-intervention becomes something people reference obliquely, if at all. It is mentioned in side conversations, in quiet acknowledgements between colleagues who shared that moment of uncertainty. It never enters official documentation. It never becomes part of institutional memory.

This is how systems protect themselves from regret.

Yet the absence of regret comes at a cost.

Without confronting the near-decision directly, the organisation misses an opportunity to examine the deeper mechanics of its own judgment. It continues to believe that the challenge lay in insufficient information rather than insufficient permission.

Permission is the harder problem.

Permission requires institutions to redefine what counts as legitimate action. It requires them to accept that acting early will sometimes look unnecessary, uncomfortable, or even wrong in retrospect. It demands tolerance for decisions whose success cannot be proven because they prevent events that never occur.

Institutions struggle with this kind of success.

Success without evidence feels indistinguishable from failure.

As a result, people internalise a quiet lesson: *wait until you can prove you were right.* By the time proof exists, the window for prevention has usually closed.

This lesson shapes behaviour long after the event has faded from view.

New scenarios arise. Different conditions, different contexts, different risks. Yet the pattern repeats. Early signals appear. They are noticed. They are discussed cautiously. They are monitored. They do not yet justify action.

People remember how the last situation resolved. They remember that waiting did not immediately lead to harm. They remember how decisive the response was once action became unavoidable.

They draw the wrong conclusion.

The conclusion is not that earlier action was impossible. It is that earlier action was unnecessary.

This belief hardens slowly, reinforced by time and by the absence of visible counterfactuals. No one can point to a disaster that did not happen and say, with certainty, *this was prevented.* The system has no language for that claim.

Over time, this belief becomes embedded in culture. It influences how people speak, how they escalate concerns, how they judge one another's decisions. Those who advocate restraint are seen as steady. Those who push early action are seen as speculative.

Speculation carries social risk.

As a result, the organisation grows increasingly skilled at managing what is already visible and increasingly hesitant to act on what is merely forming.

This imbalance does not announce itself as weakness. On the contrary, it presents as maturity. The system appears calm, measured, and controlled. It avoids unnecessary disruption. It maintains credibility.

But beneath this calm lies a narrowing of imagination.

Imagination is what allows institutions to act before certainty. It is what connects weak signals into plausible futures. Without it, systems become prisoners of the present moment, reacting only when the future has already arrived.

The subtle surprise, revealed only over time, is that the system's greatest strength–its discipline–has become the source of its vulnerability.

Discipline, when unexamined, hardens into rigidity.

Rigidity resists early movement.

And so the organisation continues forward, confident in its processes, proud of its performance, and largely unaware that the most consequential decisions are still being deferred not because they are invisible, but because they are uncomfortable to own.

The moment before the next crisis is already forming. It will not arrive loudly. It will appear as another ordinary stretch of time filled with competence, caution, and reasonable delay.

Once again, the system will see it. Once again, it will hesitate.

And once again, when action finally becomes unavoidable, the story will be told as if that was the first moment that truly mattered.

What changes next is not the environment, but how people relate to it.

A few individuals begin to notice that they are no longer surprised by uncertainty. What once prompted discussion now elicits a nod. The unfamiliar has become familiar, and familiarity carries a deceptive comfort. It suggests that the system has adapted, when in truth it has merely adjusted its expectations downward.

This adjustment is quiet, almost imperceptible. It shows up in how long people linger on certain screens, in how quickly emails are answered, in the tone of internal briefings. There is less probing, more summarising. Less curiosity, more closure.

Closure feels productive.

One afternoon, a junior analyst asks a question that feels slightly out of place. It is not a challenge, not a critique, just an observation that two indicators, when viewed together, seem to tell a different story than when viewed apart. The room listens politely. Someone responds by explaining why the indicators are assessed separately. The explanation is correct. The question dissolves.

Later, the analyst realises that the explanation answered how things are done, not why. The distinction lingers. It is a small surprise, the kind that does not interrupt work but quietly reframes it.

In another part of the organisation, a routine external inquiry arrives–nothing urgent, nothing accusatory. It asks for clarification on assumptions that underpin current posture. The request is standard. The response is drafted carefully, reviewed thoroughly, and approved quickly.

What is notable is not the content of the response, but the confidence with which it is delivered. The organisation believes what it is saying. That belief has been reinforced by months of apparent stability. Confidence becomes a substitute for proof.

The inquiry closes. No follow-up is required.

This outcome reinforces a subtle narrative: external scrutiny has arrived and found nothing wanting. The system reads this as validation, even though the inquiry never asked about the near-decisions, the moments of hesitation, or the signals that never quite aligned.

Those questions were never on the form.

As time passes, people begin to conflate *defensibility with correctness.* Decisions that can be justified are assumed to be right. The fact that a choice can be explained convincingly becomes evidence that it was the right choice.

This conflation is dangerous because it rewards articulation over imagination.

Imagination asks uncomfortable questions. It explores possibilities that cannot yet be defended. In systems that value justification, imagination learns to stay quiet.

Another surprise arrives, this time from within the system itself. A long-standing assumption is quietly revised–not because it failed, but because it was never tested under these conditions before. The revision is minor, technical, and easily absorbed.

Yet it changes the shape of the problem.

A few people notice. They sense that this adjustment, small as it is, shifts the margin more than expected. They exchange glances.

Someone suggests revisiting earlier discussions. Another suggests waiting to see if the change has any practical impact.

Waiting wins again.

The system absorbs the revision without revisiting its posture. Adaptation replaces reflection.

As weeks turn into months, the organisation becomes increasingly skilled at operating in this extended "pre-event" state. It refines routines, sharpens reporting, and streamlines communication. From the outside, it looks exemplary.

From the inside, it feels oddly suspended.

There is a sense–rarely spoken–that the organisation is holding its breath without realising it. People are busy, engaged, professional. Yet something feels unresolved, as though a question has been asked but never answered.

This unresolved quality produces another quiet effect: people stop expecting resolution. They begin to treat uncertainty as permanent rather than transitional. When uncertainty becomes permanent, action becomes exceptional.

Exceptional actions require exceptional justification.

The threshold for action rises without being formally adjusted.

One day, a simulation exercise is conducted. It is well-designed and challenging. Participants perform admirably. The exercise concludes with positive feedback and a sense of accomplishment.

What goes unremarked is that the scenario assumes the event has already occurred. The exercise does not explore how it might have been prevented. Prevention sits outside the frame.

No one points this out.

The surprise here is not oversight, but assumption: the system practices response so often that it forgets to practice anticipation.

This omission is not accidental. Anticipation is harder to rehearse. It lacks clear endpoints. It demands choices without feedback.

As this pattern deepens, the organisation's internal story grows simpler. It becomes a story of vigilance, readiness and professionalism.

It leaves little room for doubt.

Doubt, however, is where foresight lives.

The next time an early signal appears, it is processed more quickly, with less discussion. It fits into an established category. The system knows what to do with it.

This efficiency feels like progress.

Yet something is lost: the pause that once allowed people to wonder whether the category itself was still valid.

When consequence finally arrives again–inevitable, unmistakable–the organisation responds with the same competence as before. The cycle repeats.

But this time, something is different.

A few people recognise the pattern while it is unfolding. They feel a brief flash of recognition–not surprise at the event, but recognition of the path that led there. The recognition is quiet, personal, and fleeting.

There is no forum for it.

Afterward, in private conversations, someone says, "It feels like we've been here before." Another nods. The statement carries weight, not because it accuses, but because it acknowledges repetition.

Repetition is harder to dismiss than error.

Yet even now, the system's learning gravitates toward response improvements. Anticipation remains elusive.

The deepest surprise, revealed only gradually, is that the system does not need more data, better models, or faster alerts. It needs permission to act when the story is still forming, when outcomes are uncertain and when success cannot be proven.

Until that permission exists, the most important moments will continue to feel premature when they occur and obvious only after they pass.

The organisation will continue to perform well, to respond effectively, and to wonder–quietly, briefly–whether something more could have been done earlier.

Then it will move on.

At first, the consequences of waiting do not appear where anyone expects them to.

They do not arrive as catastrophe or public failure. They do not come with headlines or sirens. Instead, they surface quietly, in places the organisation does not instinctively associate with risk at all.

They appear in spreadsheets.

Someone notices that a cost assumption has been revised upward. Not dramatically–just enough to reflect "prevailing conditions." Another line item absorbs a contingency buffer that did not exist before. A timeline stretches slightly, then slightly again, justified each time by prudent caution.

None of this is alarming. It is framed as realism.

Realism, like restraint, feels responsible.

Over time, these small adjustments accumulate. Budgets that once balanced comfortably now require explanation. Forecasts include new caveats. Margins thin, not because anything has gone wrong, but because uncertainty has quietly become expensive.

This is the first point at which nature begins to touch the balance sheet.

The connection is not made explicitly. No one says that delayed intervention is now costing money. The relationship is indirect, filtered through accounting language and operational rationale. Costs are attributed to "complex conditions," "external volatility," or "market pressures."

All of these are true. None of them tell the whole story.

In a finance review, a senior leader asks why a particular provision has increased. The answer is careful and technical. It references prudent risk management, alignment with peers, and the need to remain conservative given uncertainty. The explanation is accepted.

What is not discussed is that the provision exists because uncertainty was allowed to persist.

Uncertainty, once tolerated operationally, becomes liability financially.

Elsewhere, insurance terms are revisited. Premiums adjust. Coverage language tightens. Risk categories are reclassified. The organisation remains insurable, but at a higher cost and with more conditions attached.

No single decision triggers this shift. It unfolds gradually, through reassessment rather than reaction.

This is the second surprise: ***the system pays for inaction long before it pays for failure.***

In board discussions, risk begins to feel different. It is no longer an abstract possibility managed by technical teams. It appears now as exposure—quantifiable, reportable, explainable in monetary terms.

Numbers change how people listen.

When risk becomes a financial variable, attention sharpens. Questions become more pointed. Timelines are scrutinised. Assumptions are challenged with greater urgency.

What is striking is not that this scrutiny exists, but that it arrives only now.

Earlier uncertainty did not command this level of attention because it lacked a price. Now it has one.

The organisation begins to notice that it is spending money to stay in the same place. Maintaining operational posture requires increasing investment. What once felt like stability now feels like effort.

This effort is not framed as loss. It is framed as cost of doing business.

But costs, unlike risks, must be justified repeatedly.

As capital committees review allocations, conversations shift subtly. The question is no longer "Is this likely?" but "What happens if this continues?" The future is discussed less as scenario and more as burden.

Nature, once treated as background, now appears as a line item that refuses to stay flat.

Another quiet shift occurs when external stakeholders begin asking different questions. Not about preparedness or response capability, but about sustainability. About long-term exposure. About whether current models remain viable under persistent conditions.

These questions are harder to answer.

Preparedness demonstrates competence. Sustainability demands foresight.

The organisation realises, slowly, that it can no longer afford to simply wait. Not because disaster is imminent, but because waiting has become expensive.

This is the moment where the logic of anticipation collides with the logic of finance. Early action, once dismissed as speculative, now appears comparatively cheap.

The irony is unmistakable, though rarely acknowledged aloud. The system resisted acting early because justification was incomplete. It now faces pressure to act because costs are undeniable.

Financial clarity arrives where operational certainty did not.

In post-period reviews, someone observes that several investments made reactively would have been smaller, if deployed earlier. The observation is noted, but treated cautiously. Counterfactuals are difficult to prove. No one wants to speculate too openly.

Speculation is still uncomfortable.

Yet the balance sheet does not speculate. It records.

Recorded costs accumulate. They tell a story that operational narratives avoided. They reveal that delay was not neutral. It had weight. It had direction.

The organisation begins to understand something unsettling: even when nothing visibly happened, something was happening all along.

Risk was converting itself into obligation.

At this point, the language shifts again. Risk is no longer framed solely in terms of safety or continuity. It is discussed as exposure, volatility, and long-term viability. Nature is no longer external. It is embedded in financial outlooks.

This reframing changes power dynamics. Finance, governance, and strategy now enter conversations once dominated by technical assessment. Decisions accelerate not because uncertainty has resolved, but because tolerance for its cost has diminished.

The system is still reacting, but now it is reacting to economics rather than events.

This is where the story turns decisively.

Not with fire or failure, but with spreadsheets that refuse to balance and forecasts that no longer converge. Nature does not need to burn to be felt. It only needs to persist.

And persistence, over time, is more expensive than crisis.

The organisation stands at a different threshold now. Not an operational one, but a financial one. The question is no longer whether early action is justified, but whether continued delay is affordable.

This is how nature moves from the margins of awareness to the centre of institutional decision-making.

Not as an emergency.

As a balance-sheet risk.

When Nature Became a Balance-Sheet Risk

How Fire Moved from Forests to Boardrooms

At first, nothing looks like a crisis.

There are no emergency meetings. No urgent memos. No sudden shifts in posture. The organisation continues to operate much as it always has–methodically, cautiously, with a practiced confidence in its own systems. From the outside, stability appears intact.

The change begins somewhere far less visible.

It begins in spreadsheets.

A forecast is updated to reflect "prevailing conditions." A contingency buffer grows slightly larger than it was in the previous cycle. A timeline stretches–not dramatically, just enough to feel prudent. Each adjustment is minor. Each is justified. None triggers concern.

Yet together, they form the first quiet signal that nature has entered a different conversation.

Not as threat. As cost.

Finance teams are trained to notice patterns before others do–not because they are more perceptive, but because numbers refuse to remain abstract. Where operational teams can live with ambiguity, financial models demand resolution. They must converge. When they do not, something must give.

Usually, it is capital.

At a routine financial review, a senior leader pauses over a projection that looks familiar but feels subtly wrong. The numbers add up, yet the assumptions beneath them have shifted. Returns are now dependent on conditions that were once considered temporary, but are now treated as persistent.

No one objects. The assumptions are conservative. Sensible. Responsible.

What unsettles the leader is not the logic, but the speed with which this logic has become normal.

Capital, once deployed with confidence, is now hedged. Investments are staged more cautiously. Projects are broken into smaller commitments. Optionality is prized.

Optionality feels like safety. But optionality has a price.

As investment slows, execution stretches. Opportunities narrow. The organisation does not retreat–it hesitates. The hesitation is barely perceptible, but it accumulates. Momentum becomes harder to sustain.

This is not failure. It is drag.

Nature has not disrupted operations. It has begun to tax them.

Insurance is where this tax becomes explicit.

What rarely enters those discussions is that ecological damage does not reset after response.

Habitat fragmentation, species displacement, and soil degradation compound silently, long after flames are extinguished. For wildlife systems, delay is not neutral time–it is loss of recovery capacity. By the time institutions respond to fire as an asset risk, ecosystems have already absorbed costs that cannot be insured or rebuilt on quarterly cycles.

Coverage is renewed. Premiums rise. Deductibles widen. Exclusions appear where none existed before. Language tightens. The organisation remains insurable, but at increasing cost and decreasing comfort.

Insurers ask different questions now.

Not about response readiness. Not about procedures.

They ask about duration.

How long are these conditions expected to persist?

What happens if uncertainty does not resolve?

How are early signals incorporated into decision-making?

These questions are more difficult to answer than operational ones.

Preparedness demonstrates competence.

Persistence demands foresight.

Insurance does not price events alone. It prices exposure over time. As uncertainty extends, risk becomes actuarial rather than hypothetical.

The organisation realises it is paying not for what has happened, but for what might continue.

This is a profound shift.

Risk, once episodic, is now treated as structural.

Structural risk does not sit comfortably with insurance. It migrates inward. It becomes something the organisation is expected to manage–not absorb, not transfer, but own.

As insurance costs rise, the balance sheet tightens further. Provisions increase. Liquidity assumptions are revisited. What once felt like resilience now requires explanation.

In capital committees, conversations grow more serious. The language shifts from likelihood to endurance. Leaders ask not whether the organisation can withstand an event, but whether it can sustain prolonged exposure without eroding value.

Time becomes the dominant variable.

And time, unlike incidents, cannot be mitigated with response alone.

Regulatory attention follows not because of any single failure, but because patterns persist. Regulators are drawn not to crises, but to trends that suggest systemic exposure.

Requests for information arrive. They are measured, formal, and procedural. They ask how early indicators are evaluated, how decisions are escalated under uncertainty, and how accountability is maintained when action is deferred.

The organisation responds confidently. Processes exist. Decisions are documented. Governance frameworks are robust.

And yet, the questions continue.

Not because regulators doubt competence, but because they sense misalignment. They see systems optimised for reaction operating in an environment that demands anticipation.

Governance grows heavier.

Heavier governance slows action. It adds review layers. It demands justification not only for what is done, but for what is not done. Silence must now be explained.

Explaining inaction is harder than explaining action.

In explaining inaction, assumptions are exposed. Assumptions that were once implicit are now scrutinised. Why was this not escalated earlier? What criteria delayed intervention? Who held authority in the absence of certainty?

These questions do not accuse. They clarify.

Clarification is destabilising.

As governance tightens, leaders become more cautious not because they lack confidence, but because every decision now carries legal, financial, and reputational weight. Acting early, once merely uncomfortable, now appears risky in a different way.

The paradox sharpens: the more risk persists, the harder early action becomes.

Investor logic introduces the final pressure.

Investors do not ask whether the organisation is safe. They ask whether it is viable. They are less interested in preparedness narratives than in structural coherence. They want to understand how uncertainty translates into return, how exposure affects valuation, and how long ambiguity can persist before confidence erodes.

Investors listen carefully for alignment.

Alignment between what the organisation knows, what it authorises, and what it funds.

Where alignment is absent, capital hesitates.

This hesitation does not arrive as withdrawal. It arrives as caution. Valuations soften. Expectations adjust. Growth assumptions are revised. The cost of capital edges upward.

In one meeting, a decision-maker pauses longer than usual. The numbers make sense, yet the assumptions beneath them feel newly fragile. What was once treated as temporary is now embedded in projections as persistent.

The question that forms is not whether the risk exists, but whether acting early can be defended if nothing visibly happens.

The pause passes. The meeting moves on.

Capital adjusts quietly. Authority waits.

No one panics. But everyone notices.

At this point, something fundamental shifts. The organisation realises it is being judged not on outcomes, but on logic. Not on how well it responds, but on how coherently it governs uncertainty.

Data alone is no longer enough. Prediction without authority does not reassure markets. Governance without legitimacy slows decision-making. Capital without confidence demands a premium.

The threads begin to converge.

The organisation recognises slowly and reluctantly that it has treated these domains separately. Data belonged to analysts. Decisions belonged to leaders. Capital belonged to finance.

Nature has forced them into the same room.

This is the approach to the climax.

Not through catastrophe, but through convergence.

In one meeting, operational leader explain why current posture remains technically sound. Immediately afterward, finance leaders present projections showing continued margin erosion under persistent conditions. Governance leaders outline increasing regulatory expectations. Investor relations describe growing scepticism around long-term exposure.

Each presentation is correct.

Together, they contradict one another.

The silence that follows is different from earlier silences. It is not hesitation. It is recognition.

Recognition that the organisation is managing risk operationally while absorbing it financially and justifying it institutionally.

Recognition that waiting is no longer neutral.

Delay now has direction.

The most unsettling question finally surfaces–not as accusation, but as inevitability: If we continue as we are, then what exactly are we optimising for?

No one answers immediately.

Because the answer is uncomfortable.

The organisation realises it has been optimising for explainability rather than resilience. For defensibility rather than anticipation. For decisions that look right at the time rather than decisions that change trajectories.

Somewhere else, the earlier note still exists–not on a screen, not in a report, but in memory. It has aged into certainty too late to be useful. What once felt premature now feels inevitable.

This is the true climax.

Not the arrival of disaster, but the collapse of a belief that waiting for certainty is the safest position.

In this moment, something new becomes possible.

The organisation begins to entertain the idea that authority might come not from proof, but from pattern. That legitimacy might be earned not by waiting, but by explaining why early action is rational under uncertainty. That capital might flow more confidently when decision logic is coherent, even if outcomes remain uncertain.

This idea is fragile. It challenges decades of institutional conditioning. It asks leaders to accept that acting early may look wrong before it looks right.

But it also offers release.

Release from the paralysis of waiting.

By the end of this section, the organisation stands at a threshold unlike any it has monitored before. This threshold is not environmental or operational. It is philosophical.

It must decide whether it is willing to act before nature forces alignment through cost, regulation, and capital pressure.

Nature has not burned the balance sheet.

It has bent it slowly, persistently, and predictably.

Chapter 1 shows how modern institutions, built to react to certainty, absorbed risk quietly until it surfaced as cost, governance pressure, and capital erosion.

Chapter 2 begins at the point where this logic breaks–introducing an integrated approach that recognises patterns earlier, legitimises action under uncertainty, and restores financial confidence before consequence forces alignment.

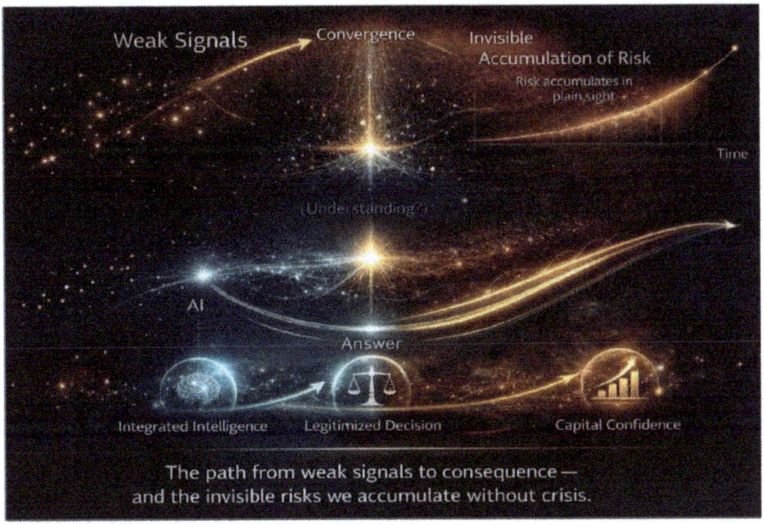

Figure 1.1 Risk does not explode - it accumulates, unseen, until it cannot be ignored.

Chapter 2 begins at the moment this bending can no longer be ignored.

It begins with a different question:

If institutions struggle to legitimise early action, then what happens when systems emerge that can see convergence before humans agree it exists?

And will institutions be ready-not just to build those systems, but to trust them?

The fracture between how risk is perceived and how it is paid for becomes impossible to ignore. Operational teams still speak the language of readiness and monitoring. Governance teams speak of defensibility and process integrity. Finance speaks of exposure, erosion, and conditional commitment.

Each group is correct in isolation.

Together, they are incompatible.

This incompatibility does not announce itself as failure. It appears instead as friction-decisions that take longer, approvals that require more explanation, investments that hesitate at the final step.

Momentum slows not because people disagree, but because no single logic can reconcile all concerns at once.

The organisation begins to experience what can only be described as decision drag.

Every major choice now requires navigating multiple, partially conflicting truths:

- The risk has not yet materialised.
- The cost of waiting is already real.
- Acting early is hard to justify.
- Acting late is becoming unaffordable.

There is no procedural answer to this tension.

This is where the old model begins to collapse–not through breakdown, but through overload.

In one extended leadership session, the conversation circles repeatedly around the same point. Someone asks how much confidence is required before committing resources. Someone else asks how much cost can be tolerated before confidence becomes irrelevant. The room recognises the circularity but cannot escape it.

This is the emotional centre of the climax.

The organisation realises that it has been using certainty as a substitute for trust.

Certainty justified action because it removed doubt. But in a world where risk accumulates quietly and persists indefinitely, certainty arrives too late to be useful. By the time certainty exists, the balance sheet has already absorbed the impact.

The uncomfortable realisation follows: the organisation does not need more certainty–it needs legitimacy earlier.

Legitimacy to act when signals converge.

Legitimacy to intervene before proof becomes loss.

Legitimacy to explain decisions made under uncertainty, rather than apologise for delays made in its name.

This is not a technical problem. It is a governance problem.

And beneath that, it is a trust problem.

Trust between data and decision-makers. Trust between institutions and regulators. Trust between organisations and capital.

Without trust, action waits for certainty. With trust, action can precede it.

The organisation begins to see, perhaps for the first time, that prediction alone was never the missing piece. Data existed. Signals existed. Models improved steadily. None of that translated into early action because ***prediction without authority has no force.***

Authority, in the old system, flowed only after thresholds were crossed. But thresholds were designed for episodic risk, not persistent convergence.

What the organisation faces now is not a sharper threshold, but a gradient–a slow, sloping increase in consequence that never triggers alarms, yet steadily drains resilience.

This is where capital becomes the final arbiter.

Capital does not wait for certainty. It reacts to coherence.

When decision logic is fragmented, capital demands a premium. When governance hesitates, capital pauses. When risk is visible but unmanaged, capital withdraws quietly–not in protest, but in prudence.

The organisation realises that it is being priced not on its exposure, but on its decision maturity.

This recognition lands heavily.

Because it means the organisation is no longer being judged on whether it can respond to crisis, but on whether it can ***act before crisis becomes the only justification left.***

At this point, the old narrative–that waiting is responsible–can no longer be sustained. Waiting has become the most expensive option.

The climax is not a decision. It is an inversion.

The organisation finally understands that early action is not reckless–it is *cheaper, clearer, and more defensible* when supported by transparent logic.

And that logic must do three things simultaneously:

- ***See patterns early***, before thresholds are crossed.
- ***Explain those patterns clearly,*** so action is legitimate across institutions.
- ***Give capital confidence***, so resources move before cost forces alignment.

Until now, these three requirements lived in different rooms.

Now they must coexist.

This is the moment Chapter 1 ends–not with resolution, but with inevitability.

The organisation stands at a boundary where reaction has reached its limit. Beyond this point, responding better will no longer compensate for acting late. The only way forward is to change *how authority is granted*–from certainty-based to convergence-based.

Chapter 2 begins precisely here.

Not with technology as saviour. Not with models as solutions.

But with a different proposition entirely:

What if intelligence could surface convergence early, explain it in human terms, and make action legitimate *before* capital begins to retreat?

What if systems could earn trust faster than events earn permission?

And what if institutions were finally ready to act–not because they were forced to, but because they could justify it in time?

That is where the story must go next.

What This Chapter Established

The Failure Before the Fire

Wildfires and ecosystem collapse are rarely sudden. They emerge as early signals and persistent patterns that institutions increasingly detect–but fail to act on. The problem was not a lack of data or foresight, but an inability to convert early risk into legitimate action. In fire prevention and wildlife protection, waiting felt responsible–until the cost of waiting became invisible.

Executive Takeaways — What This Means for Leaders

- Disasters no longer begin with visible events. They begin with patterns that remain below decision thresholds.
- Early signals are rarely absent—they are simply not granted authority.
- Waiting is not a neutral act. It is a decision to allow conditions to converge.
- Institutions are not failing to detect risk. They are failing to act before justification becomes undeniable.
- The moment of responsibility is no longer when damage occurs. It is when persistence becomes visible.

What Comes Next

How Risk Speaks Before It Burns

If fire and ecological loss announce themselves early, then prevention depends on understanding how risk communicates. The next chapter examines the language of risk and how AI begins to perceive danger long before smoke appears.

Chapter 2
The Language of Risk

How Danger Speaks Long Before It Burns

Chapter 2 confronts a quiet but consequential truth: modern risk rarely announces itself as an event. It emerges instead as a pattern–persistent, distributed, and often indistinguishable from normal variation until it is too late to respond freely.

Most institutions are not blind to danger. They are trained to recognise it in the wrong form.

This chapter examines how capable organisations equipped with data, expertise, and good intentions still fail to act early, not because signals are absent, but because those signals never achieve legitimacy within existing decision structures.

At the centre of this chapter is a simple reframing: risk is not a spike to be detected, but a language to be learned.

That language speaks through duration rather than intensity, recurrence rather than novelty, and convergence rather than singular causality. Human judgement is well suited to noticing anomalies. It is far less effective at holding persistent unease across long time horizons without institutional support.

This is not a failure of people. It is a limitation of system design.

Disasters Start as Patterns

The first section shows how early danger appears–not as alarm, but as misfit.

Conditions linger longer than expected. Small anomalies recur. Workarounds quietly increase. Systems remain operational, compliant, and defensible. Teams adapt professionally, masking degradation rather than exposing it. Over time, endurance is mistaken for safety.

Thresholds remain unbroken.

Alerts clear themselves.

Governance waits.

The institution does not ignore risk. It accommodates it.

This section demonstrates why persistence is the most dangerous signal of all and why organisations trained to respond to events struggle to legitimise action when nothing has yet "gone wrong."

Machines That Notice First

The second section introduces a pivotal shift: machine perception changes what counts as evidence.

Unlike humans, machines do not forget. They do not normalise repetition. They retain memory across time, noticing patterns that stretch beyond attention spans, organisational boundaries, and reporting cycles. What humans experience as vague unease, machines preserve as continuity.

This does not make machines wiser than people. It makes them structurally different.

As machine memory accumulates, evidence evolves. It no longer requires consequence to justify concern. Duration itself becomes meaningful. Persistence gains weight. Near-interventions become possible not because catastrophe is predicted, but because convergence is visible.

This section also explores why institutions resist this shift. Early action feels illegitimate without events to defend it. Audit frameworks, legal accountability, and capital governance are still calibrated to episodic risk. Acting before harm demands a new kind of justification–one that most systems were never designed to support.

When Signals Converge

The third section shows what happens when patterns begin to align across domains.

Environmental conditions, infrastructure strain, human adaptation, financial exposure, regulatory pressure, and market signals start telling the same story–independently, but simultaneously. No single signal dominates. Together, they remove ambiguity.

This convergence creates pressure the institution cannot ignore. Not urgency, but obligation.

At this point, the problem is no longer detection. It is coordination.

The organisation begins to recognise that early intelligence, governance legitimacy, and capital sustainability are not separate concerns to be handled sequentially. They are interdependent conditions that must be designed to work together or anticipation collapses.

This is where the architecture later formalised as L-ARC™ begins to surface not as a framework imposed from theory, but as a necessity discovered through lived constraint.

Chapter 2 establishes a critical insight for leaders:

Waiting is not neutral. It is a decision with compounding cost.

Institutions often believe restraint is the ethical choice until evidence becomes undeniable. This chapter demonstrates that, under conditions of persistent risk, waiting is itself an action–one that narrows future options, shifts capital under pressure, and converts anticipation into reaction.

The chapter reframes responsibility. It shows that legitimacy does not begin after harm occurs. It must be constructed early enough to matter.

Why This Chapter Matters for Practitioners?

- For executives, boards, regulators, and risk leaders, Chapter 2 provides:
- A new vocabulary for early danger that does not rely on crisis
- A defensible explanation for why competent organisations still acts too late
- A reframing of evidence that aligns with governance and audit realities
- A foundation for acting early without requiring prediction certainty

Most importantly, it prepares the reader for what follows.

Once an institution learns to hear the language of risk, it cannot unhear it. Awareness becomes cumulative. Responsibility expands upstream. The question shifts from "Is there enough evidence to act?" to "What kind of evidence arrives early enough to matter?"

Chapter 2 ends with that question fully articulated.

The chapters ahead focus on what it takes to answer it–operationally, institutionally, and ethically–without losing legitimacy, trust, or capital discipline.

Disasters Start as Patterns

Weak Signals, Anomalies and Quiet Warnings

The watcher is still there.

They have not moved roles. They have not gained authority. The screens look much the same as they always have. Alerts still clear themselves. Systems still behave, mostly.

What has changed is not what they see. It is how long they have been seeing it.

The pattern that once felt like an anomaly no longer resolves. It stretches across days, then weeks. It does not intensify dramatically. It persists. It hovers just below the level that would demand escalation, as if deliberately avoiding legitimacy.

The watcher no longer marks the moment privately. There have been too many of them. Instead, they develop a sense for how the system behaves *when it is slowly drifting away from its own assumptions.*

This is not something the system is built to recognise.

Most organisational intelligence is tuned to disruption. Spikes, failures, breaches, and thresholds are easy to name. They fit the grammar of response. They justify authority and activate protocol.

Patterns do not.

Patterns are uncomfortable because they do not offer a single point of causality. They unfold across time rather than space. They do not tell a clear story. They simply repeat–slightly differently each time–until repetition itself becomes the signal.

The institution does not treat repetition as danger. It treats it as experience.

In fire-prone landscapes, these patterns are not theoretical. They appear as drying fuel corridors, altered animal movement, stressed vegetation, and infrastructure heat signatures that converge weeks or months before ignition. AI does not predict fire here–it remembers how ecosystems behave when stability is eroding.

Experience, after all, teaches what can be endured.

And endurance is often mistaken for safety.

The watcher notices this shift first, not because they are more intelligent, but because they are closer to continuity. Their work requires them to notice what *normally happens when nothing is wrong*. When that "normal" begins to stretch, the stretch registers–not as alarm, but as misfit.

They try, once, to articulate it.

The language fails them.

There is no sentence that captures persistence without sounding speculative. No metric that proves drift without invoking hindsight. No chart that makes a case for unease.

What exists instead is a collection of small facts:

- something lasting longer than it should
- something recurring more often than before
- something interacting with something else in a way it did not previously

Each fact alone is unremarkable.

Together, they suggest direction. Direction is dangerous.

Direction implies trajectory, and trajectory implies responsibility. Institutions are cautious with responsibility that cannot yet be defended.

So, the watcher does what professionals are trained to do: they adapt. They adjust expectations. They incorporate the pattern into their mental model of "how things are right now."

Adaptation is rewarded. Escalation is not.

Across the organisation, similar adjustments are taking place.

A team revises its working assumptions. Another builds additional buffer into planning. A third develops informal workarounds that smooth friction without raising questions. None of these actions are wrong. All of them are reasonable.

Collectively, they form a system that is coping, not seeing.

This is the first deep irony of modern risk: the better institutions are at coping, the longer they can avoid recognising that something fundamental has changed.

Coping creates the illusion of control.

The institution still measures success in terms of continuity. As long as operations continue, as long as outputs are delivered, as long as no threshold is breached, the system believes it is functioning as intended.

But continuity is not the same as stability.

The difference matters.

Patterns live in that difference.

The watcher remembers.

But memory is not evidence.

In a meeting far removed from the screens, someone mentions that recent conditions resemble an earlier period. The resemblance is not exact. The context is different. The comment is acknowledged politely and set aside.

Institutions are uncomfortable with resemblance. Resemblance suggests analogy, and analogy is a weak foundation for action.

Action prefers certainty.

So, the institution waits.

What it does not realise is that waiting is not neutral. Waiting allows patterns to deepen, interactions to multiply, and consequences to entangle themselves across systems.

By the time the pattern becomes undeniable, it will no longer be simple.

This is the second irony: patterns are easiest to act on when they are hardest to prove.

Early, patterns are shallow but separable. Later, they are deep but intertwined.

The institution has been trained to act late.

The watcher sees this, dimly.

They see how anomalies are treated as noise, how repetition dulls attention, how persistence becomes background. They also see their own limits. There is no single moment they can point to and say, "This is it."

Instead, there is accumulation.

Accumulation does not trigger alarms.

It triggers fatigue.

Fatigue spreads quietly. Not as panic, but as resignation. People begin to speak of conditions as "the way things are now." This phrase is rarely examined. It carries with it an implicit surrender of expectation.

Once expectation is surrendered, escalation becomes almost unthinkable. Why escalate something that has already been accepted?

This is how disasters begin.

Not with denial, but with accommodation.

The watcher remains at their post, watching the system behave increasingly like something it was never designed to be. They are not afraid. They are not surprised. They are uneasy in a way that has no outlet.

They are watching a language unfold–one that speaks in duration, recurrence, and interaction–and they know the institution does not yet understand how to listen.

That understanding, when it comes, will not come from them.

It will come from something that does not forget, does not normalise, and does not grow tired of repetition.

And that is where the chapter must go next.

The institution begins to develop a strange confidence.

Not confidence in outcomes, but confidence in endurance.

Each time the system absorbs another deviation without visible consequence, it reinforces the belief that it can continue to do so. Each time a near-miss resolves quietly, it becomes evidence that escalation would have been premature. Over time, restraint becomes not just a habit, but a virtue.

This is how caution transforms into inertia.

In one part of the organisation, a routine review quietly changes tone. The agenda is the same. The metrics are familiar. But the discussion lingers longer than usual on items that once passed quickly. People begin to speak in qualifiers: *so far, for now, within limits*. None of this is alarming. All of it is new.

Someone asks whether the pattern is seasonal.

Someone else suggests it reflects increased utilisation.

A third points out that conditions are different this year.

Each explanation is plausible. Each one lowers the perceived urgency. The meeting ends with no decision, because no decision feels necessary.

What no one says aloud is that the explanations are beginning to stack rather than resolve.

Stacking explanations is a warning sign. It suggests that no single story fully accounts for what is happening. But institutions prefer multiple partial explanations to one unsettling question.

The watcher notices the change in conversation even when they are not present. It shows up later, in summaries and second-hand remarks. The language has softened. Certainty has been replaced with manageability. The system is no longer trying to understand what is happening; it is trying to remain comfortable with not fully understanding it.

Comfort is dangerous.

Comfort delays curiosity.

A second vignette unfolds away from screens and dashboards, in a policy setting designed precisely to manage uncertainty.

A working group is convened to review operating assumptions. The group is experienced. Its members have lived through disruptions before. They know the cost of overreaction, and they carry the institutional memory of false alarms that once led to unnecessary expense and embarrassment.

They approach the review carefully.

The data is presented conservatively. Trends are shown with caveats. Variability is acknowledged but contextualised. The group agrees that nothing in the material suggests immediate action.

Someone asks whether the persistence itself should be considered.

The question hangs briefly.

Persistence is harder to operationalise. It does not fit neatly into existing criteria. It lacks a clear response pathway. The group agrees to monitor.

Monitoring feels responsible.

No one realises that they have just formalised waiting.

As weeks pass, monitoring becomes the dominant mode of engagement. Dashboards are refined. Reports are requested more frequently. Reviews are scheduled closer together. All of this activity creates the impression of vigilance.

Vigilance is mistaken for anticipation.

In reality, the system is still waiting for permission to act–permission that only arrives when something crosses a predefined line.

The lines remain uncrossed.

Thresholds, once invisible, now quietly govern everything.

They determine what language can be used, what actions can be justified, and what concerns can survive escalation. As long as metrics remain on the "right" side of the line, restraint feels mandatory. To act early would require explaining why thresholds no longer matter.

That explanation does not yet exist.

Thresholds were designed for a different kind of risk–episodic, sudden, and bounded. They are excellent at detecting spikes. They are poor at detecting drift.

Drift does not confront thresholds. It evades them.

The system watches conditions approach the line, hover near it, then retreat slightly–not enough to resolve, but enough to avoid consequence. Each retreat reassures. Each approach is explained as fluctuation.

Over time, the organisation begins to treat proximity to thresholds as normal.

Normalisation is the final stage of blindness.

A third vignette takes place in the field, far from governance discussions and capital models.

A team adjusts its practices again. The environment is behaving differently, but not violently. Equipment degrades faster. People tire sooner. Small failures become more frequent, but remain manageable.

The team compensates through skill and experience. They plan more carefully. They move more slowly. They share knowledge informally about what works now.

Their professionalism is impressive.

It is also masking risk.

The reports they submit remain measured. They do not exaggerate. They do not speculate. They describe what they see, not what they feel.

What they feel is that margins are thinner than they used to be.

There is no field in the report for that.

The organisation's centre remains calm.

No incident has occurred. No rule has been broken. No authority has been challenged. From the outside, everything appears controlled.

This is when the near-intervention moment arrives.

It does not announce itself dramatically. It emerges as a proposal–modest, reversible, carefully framed. The proposal does not claim catastrophe. It does not predict failure. It suggests adjustment based on observed convergence.

The language is cautious. The scope is limited. The cost is contained.

It is, by all reasonable standards, prudent.

The discussion that follows is thoughtful.

Questions are asked about precedent. About proportionality. About the risk of acting without visible justification. No one opposes the proposal outright. No one advocates strongly for it either.

The room is balanced.

Balance feels safe.

But balance also dilutes momentum.

The proposal is neither approved nor rejected. It is deferred–pending further clarity, additional data, or a more definitive signal.

The deferral feels sensible.

It is also decisive.

Because what the institution has decided, without naming it, is that patterns are still not enough.

The watcher hears about the deferral later, indirectly. It confirms what they already suspected: the system is waiting for an event.

Waiting for an event feels rational.

Events resolve ambiguity.

But events also arrive with cost.

The watcher understands now that the institution is trapped between two fears: the fear of acting too early, and the fear of explaining why nothing happened if it did.

The first fear always wins.

Because it is immediate, personal, and reputational.

The second fear is abstract–until it is not.

This is the quiet failure of thresholds.

They do not fail dramatically. They do not collapse. They continue to function exactly as designed. That is the problem.

They function in a world where risk announces itself loudly, crosses lines decisively, and demands response. They do not function in a world where risk accumulates gradually, interacts across domains, and reshapes baselines without ever declaring itself.

The organisation begins to feel the consequences of this mismatch, not as crisis, but as constraint. Options narrow. Flexibility diminishes. Costs creep upward in ways that are difficult to attribute to any single cause.

Attribution matters. Without attribution, learning stalls.

The system senses that something has been missed, but cannot yet say what.

The climax of this section is not an explosion.

It is a recognition.

The organisation realises–not intellectually, but viscerally–that it has been asking the wrong question all along.

The question was never: *Is there enough evidence to act?*

The question should have been: *What kind of evidence arrives early enough to matter?*

That shift changes everything.

Because it reveals that the system has been optimised for justification, not perception. It has been designed to respond convincingly, not to listen carefully. It waits for danger to become legible in its existing language.

But danger is speaking a different language now.

One of duration.

One of recurrence.

One of interaction.

A language the institution hears, but does not yet understand how to trust. The watcher understands something else, now, too.

They understand that their unease was never the problem.

The problem was that unease had no institutional form.

There was no place for it to live, no process that could hold it, no authority that could act on it without apology.

Human perception reached its limit not because it was wrong, but because it was alone. The organisation does not realise it has crossed a line.

That is the nature of the line it has crossed.

It is not a boundary marked by consequence, but by posture. At some point, the organisation stops asking whether the pattern represents emerging danger and begins asking how long it can operate under these conditions. The difference is subtle but profound. One question is about understanding. The other is about endurance.

Endurance becomes the organising principle.

Meetings that once focused on interpretation now focus on mitigation. Language shifts from *why* to *how*. How do we schedule differently? How do we allocate resources more efficiently? How do we avoid fatigue, bottlenecks, reputational exposure?

These are competent questions.

They are also downstream questions.

Downstream of a decision that has already been made without being named: the decision to accept the pattern as a feature of the present rather than a warning about the future.

A new vignette emerges quietly inside a compliance review.

The review is not prompted by failure. It is routine, cyclical, designed to ensure alignment with standards and expectations. The auditors are thorough. They ask for evidence, documentation, traceability. The organisation provides all of it. Controls are in place. Processes are followed. Exceptions are explained.

The review passes.

What the review does not capture is the slow migration of assumptions. The controls are calibrated against thresholds that remain technically unbroken. The documentation reflects decisions that are defensible in isolation.

Compliance confirms legitimacy. Legitimacy reinforces waiting.

The watcher hears about the review outcome later. It changes nothing. It confirms everything.

They have learned, by now, that formal validation often lags reality. Validation is backward-looking. It assesses whether rules were followed, not whether rules still make sense.

The watcher is not frustrated. They are resigned.

Resignation is more dangerous than frustration, because it does not demand release.

Another vignette unfolds at the intersection of institutions.

A coordination call is held between agencies. The purpose is alignment, not alarm. Information is shared cautiously. Each party describes its own view of conditions, framed by its own mandates and constraints.

The descriptions overlap, but do not fully align. One agency sees strain. Another sees variability. A third sees elevated but manageable exposure. No one sees cause for escalation.

No one sees the whole.

Each agency leaves reassured that others are coping too.

This reassurance is collective.

It is also false.

The organisation's internal reporting grows more sophisticated. Dashboards multiply. Visualisations improve. Trends are smoothed, contextualised, normalised. Each improvement increases clarity locally while further fragmenting global understanding.

Clarity without synthesis becomes noise.

People begin to joke quietly about how much data is now available and how little it seems to settle anything. The joke carries a trace of unease, but humour absorbs it.

Humour is another coping mechanism.

The near-intervention moment returns, differently shaped this time.

A revised proposal circulates. It is narrower than before. The language is even more cautious. It emphasises reversibility, optionality, learning. It avoids words like risk and threat. It speaks instead of *insight, preparation, positioning*.

The proposal is stronger. It is also weaker.

Because it has adapted itself to the institution's fear of being wrong early.

The discussion is shorter this time. The questions are fewer. The deferral is quicker.

This is progress, of a kind.

Thresholds remain uncrossed.

They have become almost irrelevant, not because they are obsolete, but because the organisation has learned how to operate indefinitely beneath them. This is a form of mastery the system was never designed to achieve.

The system has outgrown its own safeguards.

A vignette unfolds around insurance and external risk assessment.

Premiums adjust slightly. Coverage terms are reviewed. Exclusions are clarified. None of these triggers alarm internally. These are market signals, not operational ones.

But markets do not wait for certainty.

Markets respond to coherence.

The organisation is still coherent internally. Externally, the story is less clear. Risk appears persistent. Mitigation appears reactive. Preparation appears bounded by thresholds that no longer reflect reality.

Capital does not flee.

It hesitates.

Hesitation is the market's version of unease.

The watcher understands markets intuitively, even if they are far removed from them. They recognise hesitation because it mirrors what they have been feeling all along.

A pattern that does not resolve.

A signal that does not escalate.

A system that does not quite trust itself.

Another vignette emerges during a talent discussion.

Fatigue is raised as a concern. Retention is discussed. The organisation prides itself on caring for its people. Adjustments are proposed: rotations, support, incentives.

Again, these are downstream responses.

No one asks why fatigue has become structural.

Structural fatigue is not a human failing.

It is a system telling you something.

At this point, the organisation is doing everything except listening differently.

It is monitoring more.

Reporting more.

Meeting more.

Explaining more.

What it is not doing is changing how legitimacy is granted.

Legitimacy remains tied to thresholds, events, and defensible certainty. Patterns remain suspect because they do not fit that grammar.

This is the heart of the problem.

The organisation has mistaken epistemic caution for moral responsibility.

It believes that waiting is the ethical choice until evidence becomes undeniable. It does not yet recognise that waiting, under conditions of persistent risk, is itself an action–one with consequences.

The watcher understands this now.

They no longer try to articulate unease. They have learned that the institution has no place to put it.

Instead, they observe.

Observation without outlet hardens into knowledge.

Knowledge without authority hardens into inevitability.

The final vignette of this section is the quietest.

Nothing happens.

No incident. No breach. No headline.

Just a small recalibration of future plans. A shift in what is considered achievable. A narrowing of ambition that no one formally announces.

This is the cost of missed patterns.

Not catastrophe.

Contraction.

The climax of this section settles like weight rather than fire.

The organisation finally understands–not through analysis, but through lived constraint–that it has been speaking the wrong language to itself.

It has been asking for certainty in a world that offers convergence.

It has been waiting for events in a world governed by patterns.

It has been trusting thresholds in a world that moves beneath them.

Human perception has done all it can do.

Human judgment has not failed.

It has reached its limit.

The watcher's unease, once personal, now belongs to the system itself. The institution feels it everywhere: in decisions that take longer, in capital that hesitates, in people who adapt too well.

This is the moment of inevitability.

If disasters begin as patterns, and if patterns evade human institutions trained for events, then perception must change–not in who decides, but in what is seen, remembered, and made legitimate early enough to matter.

This Section ends here.

Not with danger realised.

But with danger fully articulated–at last–as a language the institution must learn to hear.

And that language will not be learned by humans alone.

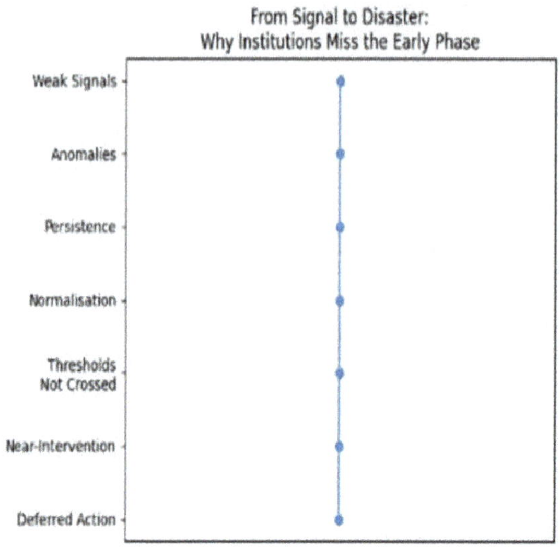

Figure 2.1 From Signal to Disaster

Disasters rarely begin with a single failure. They emerge through a sequence of weak signals, recurring anomalies, and persistent conditions that never cross formal thresholds. Institutional action is deferred not because warning signs are absent, but because legitimacy remains tied to events rather than patterns.

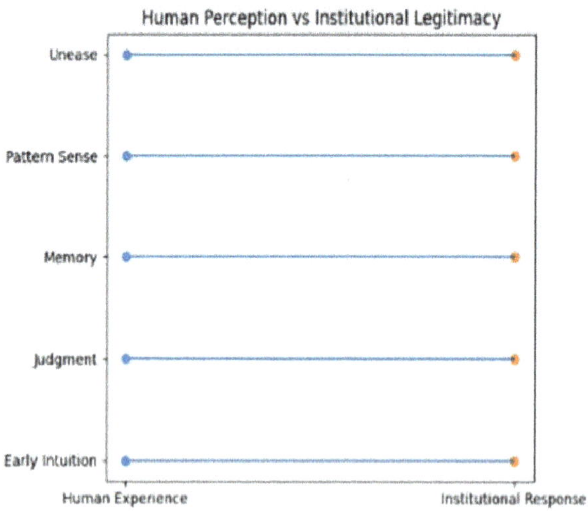

Figure 2.2 Human Perception and Institutional Response

Early risk is often perceived through unease, pattern recognition, and intuition. However, institutional legitimacy depends on thresholds, documentation, and defensible certainty. The gap between perception and permission is where early intervention most often fails.

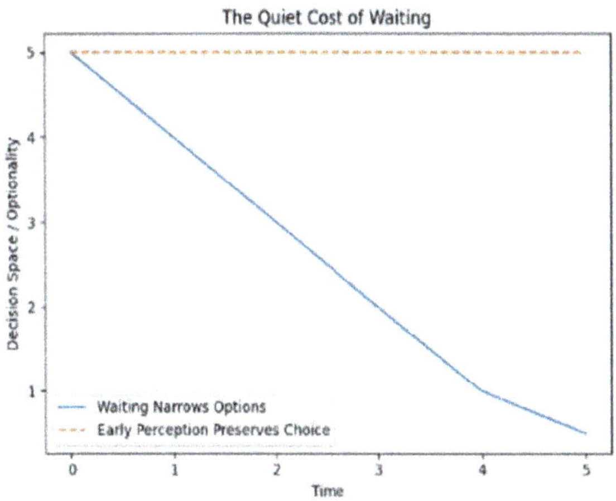

Figure 2.3 The Quiet Cost of Waiting

As time passes without intervention, the range of available decisions narrows. Waiting preserves procedural comfort but erodes strategic optionality. Early perception expands choice even when outcomes remain uncertain.

The institution does not pause after this realisation.

It cannot.

There are schedules to meet, obligations to fulfil, expectations to manage. The system absorbs insight the same way it absorbs strain– quietly, without ceremony, folding it into routine.

Understanding does not immediately change behaviour.

It changes tone.

Meetings acquire a different gravity. Words like *manageable* and *acceptable* remain in use, but they are spoken with less conviction. Plans are framed more cautiously. Commitments are hedged. Contingencies become permanent features rather than temporary safeguards.

The institution has not admitted anything explicitly.

But it has begun to behave as though the future is narrower than it once believed. This narrowing expresses itself in subtle ways.

Procurement cycles shorten, but scopes contract. Legal reviews take longer, not because risks are greater, but because justifications must be tighter. Communications teams prepare parallel narratives–one for continuity, one for explanation–even though neither is deployed.

These are not signs of panic. They are signs of quiet recalibration.

Another vignette unfolds in a legal context.

A review is conducted to assess exposure–not hypothetical exposure, but exposure as it exists under current operating assumptions. The review is thorough. Scenarios are considered. Responsibilities are mapped.

The conclusion is cautious.

No single risk appears actionable on its own.

Liability remains diffuse.

Standards have been met.

The organisation is compliant.

Compliance, once again, delivers reassurance.

But one sentence lingers longer than the rest: *"Exposure increases materially under prolonged persistence."*

The sentence is noted. It is not escalated.

Persistence still has no owner.

In a different room, communication strategy is discussed.

Not crisis communication–that would imply an event.

Instead, the conversation is about narrative stability. How does the organisation describe itself if conditions do not improve, but do not deteriorate dramatically either? How does it explain effort without outcome? Adaptation without resolution?

No one has an answer.

The absence of an answer is itself revealing.

Institutions are comfortable telling stories about action and recovery. They are less comfortable telling stories about anticipation and restraint. The language of "almost" and "not yet" does not travel well.

This makes early action politically expensive.

The watcher senses this even without attending these meetings. They feel it in the atmosphere–in the way questions are framed, in the way certainty is prized even when it is artificial.

A further vignette emerges at the boundary between organisation and oversight.

Briefings are prepared for external stakeholders. They are accurate, careful, and unprovocative. They emphasise capability, readiness, and resilience. They avoid speculation.

Speculation, after all, invites scrutiny.

The briefings are well received.

No one asks why readiness seems to be doing so much work.

Inside the organisation, a quiet division begins to form–not between people, but between temporal perspectives.

Some focus on the present, ensuring continuity.

Others, fewer in number, focus on the future, sensing constraint.

The gap between these perspectives is not hostile. It is invisible.

Invisible gaps are the hardest to bridge.

The watcher has stopped trying to bridge it.

They have learned that language alone is insufficient. Without a structure that can hold patterns across time and domains, every attempt dissolves into polite acknowledgment.

Instead, they observe something else now: **how often the same conversations repeat with different details.**

Repetition is a form of signal.

The organisation treats it as routine.

Another vignette unfolds in finance.

Budget forecasts are adjusted incrementally. Each adjustment is justified. None of them triggers concern. Over time, however, the adjustments compound. Flexibility erodes quietly.

The question is no longer whether the organisation can afford intervention. It is whether it can afford *not* to intervene and still explain itself later.

That explanation, too, lacks language.

At this stage, the system is caught in a paradox.

It knows more than it can act on.

It sees more than it can legitimise.

It understands more than it can explain.

This is not ignorance. It is institutional saturation.

The climax of this extended arc does not arrive as insight.

It arrives as silence.

In a meeting that would once have generated debate, there is none. The options are presented. The risks are acknowledged. The constraints are understood.

No one speaks first.

The silence is brief, but it carries weight.

Silence, here, is not uncertainty. It is recognition.

Recognition that the system has reached the boundary of its current logic. Recognition that further refinement will not produce clarity. Recognition that waiting has ceased to be neutral.

This is the point at which something must change–not because disaster is imminent, but because *listening differently* has become the only responsible option left.

The institution does not yet know what that change looks like.

But it knows what it cannot continue doing.

It cannot continue relying on human memory to hold patterns across months.

It cannot continue normalising persistence as experience.

It cannot continue asking thresholds to detect drift.

And it cannot continue pretending that anticipation without legitimacy is reckless, while reaction under constraint is responsible.

The watcher understands this without articulating it.

They understand it because they have lived inside the gap between seeing and acting for too long.

What is needed now is not a better argument.

It is a different form of perception.

One that does not forget.

One that does not tire.

One that does not require permission from events to speak clearly.

This is where this section truly ends. Not at the edge of catastrophe.

But at the edge of a transformation in how risk is seen, remembered, and trusted.

The institution does not announce that it has reached this point.

It simply begins to behave as though certain futures are no longer plausible.

Long-range plans quietly lose ambition. Language shifts from growth to continuity, from expansion to protection. No single decision marks the change. It emerges across documents, conversations, and assumptions that are never compared side by side.

This is how strategic contraction happens without acknowledgement.

People still speak of resilience, but what they mean now is endurance. They still speak of readiness, but what they mean is the ability to cope longer. They still speak of risk management, but what they are actually doing is rationing uncertainty.

Uncertainty has become a resource to be spent carefully.

Another vignette appears at the political boundary.

A briefing is prepared for oversight bodies. It is careful, balanced, and conservative. It emphasises that nothing has crossed formal limits. It highlights preparedness, capability, and ongoing monitoring. It reassures.

The briefing is accepted.

No one asks what *monitoring* is meant to lead to.

No one asks what happens if conditions persist indefinitely.

Oversight functions are satisfied because no failure has occurred. Their mandate is to respond to breakdowns, not to prevent drift.

The institution understands this and tailors its language accordingly.

This is not deception. It is alignment.

The watcher understands alignment too.

They see how language adapts as it moves upward and outward. How unease is translated into neutrality. How persistence is softened into variability. How duration disappears into phrasing.

They recognise that the system is protecting itself–not from danger, but from being asked to act before it can justify itself.

Justification has become the highest currency.

A further vignette unfolds around ethics and responsibility.

Someone raises a question, gently, about long-term exposure. Not liability in the legal sense, but responsibility in the moral one. What does it mean to continue operating under conditions that are known to be changing, even if no single harm can be attributed yet?

The question is acknowledged.

It is also redirected.

Ethical frameworks, like governance frameworks, prefer clarity. They work best when cause and effect are visible. They struggle with slow accumulation and distributed consequence.

The discussion moves on.

Responsibility, once again, has nowhere to land.

Time stretches.

This is perhaps the most important shift of all.

When patterns persist without consequence, time stops behaving like urgency and starts behaving like background. Days blur. Weeks pass unnoticed. Reviews repeat with minor variations. The extraordinary becomes ordinary.

The organisation adapts its internal tempo to match this new normal. Urgency would feel inappropriate now.

Urgency would suggest danger.

Danger would demand action.

Another vignette emerges in procurement and infrastructure planning.

Replacement cycles are adjusted. Maintenance schedules are tightened. Capital expenditure is re-sequenced. Everything remains within policy. Everything is justified.

Yet the logic has changed.

Assets are no longer planned around expected conditions, but around tolerated ones.

Tolerance is not the same as design.

Design anticipates. Tolerance endures.

The watcher senses this shift most acutely.

They notice how the system now assumes persistence rather than resolution. How conversations no longer ask *when things return to baseline*, but *how long baseline can remain displaced.*

This is a profound psychological shift.

It means the institution has accepted that the world has changed–without yet accepting that its decision logic must change too.

Acceptance without adaptation is dangerous.

Another vignette unfolds around public narrative.

Communications teams discuss messaging not because something has gone wrong, but because nothing has gone right either. How does one communicate sustained effort without visible outcome? How does one explain vigilance without escalation?

There is no satisfying answer.

Public narratives, like institutional ones, prefer events.

The absence of events makes anticipation difficult to explain.

This makes early intervention socially expensive.

At this stage, the organisation is no longer simply missing signals.

It is actively organising itself around not acting on them.

This is not negligence.

It is a rational response to a system that punishes premature certainty more harshly than delayed action.

Delayed action can always be justified.

Premature action must be defended.

The watcher understands now that this asymmetry is the real driver of delay.

The final extension of this section deepens the inevitability.

The organisation begins to encounter second-order effects.

Not disasters, but constraints.

Partners grow cautious. Timelines slip subtly. Optional projects are shelved without explanation. Innovation slows, not because imagination has failed, but because risk appetite has narrowed.

No one attributes these changes to the pattern itself.

They are attributed to "the environment."

The environment becomes a convenient abstraction–a way of naming constraint without naming cause.

The organisation has become exquisitely sensitive to risk.

And utterly unable to act on it early.

It sees more.

It knows more.

It understands more.

And yet, it is more constrained than ever.

Human perception has not failed.

Human judgment has not failed.

Institutional permission has failed.

Permission that arrives only after damage is not permission at all.

It is confirmation.

The watcher no longer thinks in terms of warning.

They think in terms of inevitability.

They understand that unless perception itself changes–unless patterns can be held, remembered, explained, and legitimised without waiting for consequence–the system will continue to move forward with narrowing choices.

They do not know what that new form of perception looks like.

But they know it cannot depend on human memory alone.

This is the final resting point of this Section.

It is no longer about whether disasters start as patterns.

That has been established beyond doubt.

It is about whether institutions are willing to change how they listen–before patterns harden into costs that no amount of response can undo.

Next Section follows not as a solution.

But as an answer to a question the institution can no longer avoid asking.

Case Study

When Patterns Were Still Called Weather: Black Summer (Australia, 2019–2020)

It didn't start as a crisis. It started as a pattern.

Hotter seasons. Drier landscapes. Longer fire conditions. These signals were visible across data, models, and lived experience.

But they were still treated as variation–something within the range of normal.

Then everything aligned.

Heat, wind, dryness, and ignition converged, leading to one of the largest wildfire seasons in Australia's history. Millions of hectares burned. Ecosystems were permanently altered.

The signals were not missed.

They were misinterpreted.

What looked like weather was already risk-taking shape.

Executive Takeaway:

When patterns are ignored, disasters appear sudden.

Machines That Notice First

AI As Perception, Not Prediction Alone

The watcher notices again.

They do not know why this moment feels different. Nothing on the screens has changed dramatically. No alert has fired. No threshold has been crossed. If anything, the system appears calmer than it did before–more stable, more predictable.

That is precisely what unsettles them.

Stability has become a sign of accommodation rather than resolution. Predictability has begun to feel like exhaustion.

The watcher has learned to distrust calm.

Not because calm is dangerous, but because calm, in this system, is often the by-product of having learned how to endure conditions that were never meant to persist.

They have seen this before. Many times now. The difference is not in the signal.

It is in the response.

This time, the pattern does not disappear when attention shifts.

It does not fade into background noise.

It does not resolve quietly.

This time, the system remembers.

The machine is not introduced with ceremony.

There is no announcement.

No declaration of transformation.

No promise of insight.

It arrives as an incremental capability–an enhancement, a support, a way of "augmenting situational awareness."

The language is deliberately modest.

Institutions are wary of grand claims.

The machine is framed not as a decision-maker, but as a listener.

A better listener.

At first, it does nothing remarkable.

It ingests data the institution already has. It observes signals that humans already see. It does not challenge authority. It does not recommend action. It simply watches–continuously, across timeframes that human attention cannot hold.

The watcher is barely aware of it at first.

It exists somewhere adjacent to their work, producing summaries that are rarely read closely. Another system. Another output. Another thing to be reviewed later.

And yet, something begins to shift.

The machine does not become tired.

It does not grow accustomed to repetition.

It does not normalise persistence.

When conditions repeat, it does not treat them as experience. It treats them as reinforcement.

Each recurrence strengthens a pattern rather than dulling it. This is the first fundamental difference.

Humans habituate.

Machines accumulate.

The watcher notices the difference not in a dramatic alert, but in a subtle change to how information is presented.

What was once a series of disconnected observations begins to appear as a shape.

Not a forecast.

A contour.

Duration becomes visible. Interaction becomes legible. What once felt like intuition begins to acquire structure.

The watcher does not yet trust it.

Trust takes time.

A new vignette unfolds in a review meeting.

The machine's output is included almost as an afterthought–a supplementary slide, a visualisation appended to the end of a pack. No one expects it to change the discussion.

But when the slide appears, the room becomes quieter.

Not because it predicts catastrophe. Because it refuses to forget.

The slide does not show spikes or breaches. It shows persistence. It shows how conditions that were once independent have begun to co-vary. It shows how "manageable" states have overlapped long enough to form a regime.

The machine does not use that word.

But the implication is unmistakable.

Someone asks whether the data is correct. *It is.*

Someone asks whether this has happened before. *It has.*

Someone asks why this was not visible earlier. No one answers immediately.

Because the answer is uncomfortable. It was visible. Just not *holdable.*

This is the second fundamental difference.

Humans experience time.

Machines retain it.

What humans remember as a feeling, machines preserve as structure.

The institution reacts cautiously.

It does not rush to act. It does not cede authority. It does what it always does when confronted with something new and unsettling: it interrogates the legitimacy of the signal.

Where does this come from?

How confident is it?

What assumptions does it embed?

What happens if it is wrong?

These are not hostile questions.

They are the questions of an institution that has learned to fear premature certainty.

The machine does not answer defensively.

It does not justify itself.

It simply continues to show the same pattern, every time it is asked.

The watcher begins to rely on it–not for decisions, but for confirmation.

Not confirmation of outcomes.

Confirmation of unease.

For the first time, their intuition has an external memory.

They are no longer alone.

A further vignette unfolds in governance.

The machine's output begins to appear more regularly. It is still framed conservatively. Still positioned as advisory. Still surrounded by caveats.

But something has changed.

The conversation no longer centres on whether the pattern exists.

It centres on what it means.

Meaning is a different kind of question.

Meaning invites responsibility.

This is where the true tension of Section 2.2 begins to form.

The machine has not solved the problem.

It has destabilised the old equilibrium.

By making patterns persistent, it has removed the institution's ability to forget.

By remembering relentlessly, it has shifted the burden of justification.

Now, the question is no longer: *"Why act?"*

It is: *"Why not?"*

This is a dangerous inversion.

The watcher senses this shift immediately.

They feel relief. And fear.

Relief, because the language they have been struggling to speak now exists outside them.

Fear, because the institution has not yet decided what to do with a perception it can no longer ignore.

Machines can notice first. But they cannot decide.

Not yet.

And that is where this section will continue to go. The institution does not resist the machine openly.

Open resistance would require a position, and positions require articulation. Instead, resistance takes the form of careful distance. The machine is acknowledged, consulted, referenced and quietly bracketed.

Its outputs are described as interesting.

Its patterns are described as worth monitoring.

Its persistence is described as informative but not decisive.

This language matters.

Language is how institutions domesticate unfamiliar intelligence.

The first signs of pushback appear not in argument, but in framing.

In documents, the machine's insights are moved to appendices. In meetings, its outputs are shown last. In summaries, its conclusions are softened into observations.

No one disputes the data. They dispute its standing.

Standing is the currency of institutional action.

The watcher feels this immediately.

They recognise the manoeuvre because they have lived it themselves. What could not be dismissed is being deferred. What could not be ignored is being contextualised into harmlessness.

Context is not neutral.

Context determines consequence.

A vignette unfolds in a senior review forum.

The machine has been running long enough now to establish credibility. Its outputs are consistent. Its patterns recur across datasets that were previously treated as independent. It shows persistence with a patience that human attention cannot sustain.

The watcher sits quietly as the outputs are presented.

For a moment, the room behaves differently.

People lean forward. Questions sharpen. The machine's memory makes something visible that no one in the room can unsee: the pattern is not episodic. It is structural.

Then the familiar questions arrive.

How explainable is this?

How sensitive is it to assumptions?

How would we justify acting on this if nothing happens?

These are not questions about accuracy.

They are questions about legitimacy.

This is the first real confrontation between machine perception and institutional authority.

The machine can see. But seeing is not enough.

The institution needs to be able to stand behind what it sees.

Standing requires more than confidence. It requires defensibility.

An audit process is triggered–not because something has gone wrong, but because something new has arrived.

The audit is thorough.

Inputs are reviewed. Methodologies are examined. Biases are interrogated. Governance frameworks are consulted. Controls are mapped.

The machine performs well. Too well, perhaps.

It does not fail in ways the auditors expect. It does not produce erratic outputs. It does not collapse under scrutiny. It simply continues to show the same patterns.

This creates a problem.

Audits are designed to assess compliance and correctness. They are less equipped to assess persistence.

Persistence does not violate policy. It violates comfort.

The audit report is careful. It finds no fault.

It also offers no endorsement strong enough to compel action.

The machine is deemed fit for purpose–as an analytical support tool.

Support is a safe category.

Support does not require accountability.

The watcher reads the report later.

They feel something close to disappointment, but it passes quickly. They have learned that disappointment is another form of misplaced expectation.

What matters is not what the report says.

What matters is that the machine continues to remember.

Institutional pushback intensifies subtly.

New requirements are introduced. Confidence intervals must be tighter. Visualisations must be simpler. Outputs must align more closely with existing reporting formats.

Each requirement is reasonable.

Collectively, they slow the machine down.

This is not sabotage. It is assimilation.

The near-intervention moment emerges almost accidentally.

The machine flags a convergence that the institution has not formally defined as dangerous–but that the watcher recognises instantly. It is the same uneasy alignment they have been feeling for months, now rendered unmistakable.

The signal is not dramatic. It is persistent.

The watcher escalates.

This time, the escalation is different.

It is not based on intuition. It is not based on memory.

It is based on a pattern that has been held, reinforced, and articulated by a system that does not forget.

The escalation moves faster than expected. The institution pauses.

A meeting is convened. Not a crisis meeting. A clarification meeting.

The machine's outputs are projected again. The convergence is explained carefully. No claims of catastrophe are made. No predictions are offered.

What is offered instead is a statement of persistence: these conditions have coexisted long enough to justify attention.

This time, the room does not immediately retreat into procedural caution.

This time, the question shifts.

What would it look like to act before this becomes undeniable?

The proposal that follows is tentative but real.

It suggests a small adjustment.

A temporary reconfiguration.

A reversible intervention.

The cost is limited. The scope is contained. The language is precise.

For the first time, the institution considers acting on a pattern rather than an event.

The watcher feels a quiet surge of hope.

Hope is dangerous.

The resistance comes not from disbelief, but from fear of precedent.

If the institution acts now, then what does that say about all the times it did not act before? If it legitimises machine-held patterns as evidence, what new responsibilities does it inherit?

Responsibility scales.

Institutions know this.

The discussion tightens.

Legal implications are raised. Governance pathways are examined. External scrutiny is anticipated. Someone asks how this would be explained if nothing happens.

The old fear returns.

The proposal stalls.

Not rejected. Deferred.

Almost.

The near-intervention fails by inches.

But something has changed.

The failure is no longer invisible. It is acknowledged.

Someone says, quietly, that this feels different from previous deferrals. No one disagrees.

The watcher understands now that the machine has not been introduced to solve the problem.

It has been introduced to change the terms of recognition.

The institution can no longer pretend it did not see.

It can only decide what it is willing to do with what it sees.

The climax of this movement settles slowly.

The machine has not forced action. But it has forced honesty.

Patterns are no longer invisible. They are merely inconvenient.

And inconvenience, once recognised, does not fade easily.

It continues from here toward a deeper confrontation–not between humans and machines, but between old legitimacy and new perception.

Machines notice first.

The institution now knows that.

The harder question is whether it is willing to let noticing count.

The institution does not return to equilibrium after the first near-intervention.

It tells itself, that the discussion was healthy, the scrutiny appropriate, the deferral reasonable. But something has shifted beneath the surface. The old rhythm does not quite reassert itself. The machine's outputs continue to appear, and each time they do, they carry the residue of the almost-decision that preceded them.

Memory has entered the system.

Not human memory–institutional memory.

And institutional memory behaves differently.

The machine does not escalate its language.

It does not sharpen its claims or exaggerate its signals. It simply continues to accumulate. Each new cycle reinforces what came before. Each new interaction adds coherence rather than novelty.

What changes is not the signal.

It is the institution's tolerance for ignoring it.

The watcher notices that conversations are now starting from a different place.

No one asks whether the pattern exists anymore. That question has expired. Instead, discussions open with quieter, heavier phrases: assuming this continues... given what we're seeing over time... if we accept this as persistent...

These are conditional statements.

But they mark a turning point.

They mean the pattern has crossed from speculation into premise.

A new vignette unfolds at the boundary of agencies.

A coordination forum reconvenes–not in response to crisis, but because the same questions are now being asked in different places at once. Each agency arrives with its own data, its own framing, its own caution.

The machine's outputs are shared.

Not as a directive. As a reference.

Something remarkable happens. The patterns align.

Not perfectly, not neatly–but enough to make fragmentation visible. What each agency thought was contextual variance now appears as a shared drift viewed from different angles.

No one planned this convergence.

That is what makes it unsettling.

Cross-agency pressure begins quietly.

It does not arrive as demand. It arrives as discomfort.

Discomfort at realising that deferral in one place is compounding risk elsewhere. Discomfort at recognising that no single authority owns the pattern, yet all are subject to its consequences.

This is the problem of distributed responsibility.

Machines do not solve it. They expose it.

The institution responds by tightening process.

More coordination calls. Clearer escalation pathways. Sharper definitions of scope. The aim is not to act yet, but to ensure that if action becomes necessary, then it will be defensible.

Defensibility is still the organising principle.

But the ground beneath it is shifting.

The second near-intervention emerges from convergence rather than persistence alone.

The machine flags a configuration that has not occurred before–not a new signal, but a new alignment of existing ones. Duration, interaction, and proximity have crossed into a space that feels qualitatively different.

This time, the watcher does not hesitate.

They escalate immediately.

Not because the risk is larger. But because the pattern is clearer.

The response is faster than before.

A broader group is convened. Multiple agencies attend. Legal, operational, and financial perspectives are present. The machine's outputs are central, not supplementary.

The conversation is tense.

Not adversarial.

Honest.

Someone says what has not yet been said aloud: If we do not treat this as evidence, then what exactly are we waiting for?

The room does not answer immediately.

This is the moment where machine memory begins to change the nature of evidence.

Evidence is no longer defined by thresholds crossed.

It is defined by patterns sustained.

The machine does not claim certainty. It claims continuity.

Continuity is harder to dismiss.

Audit challenges resurface–sharper this time.

If the institution acts, then how will it justify relying on a system that "sees" in ways humans cannot fully interrogate intuitively? How will it explain causality, accountability, responsibility?

The machine does not resolve these questions.

But it reframes them.

The question is no longer whether the machine is right.

The question is whether the institution is willing to accept that memory itself can be a form of evidence.

This is a profound shift.

The proposal that follows is stronger than the first.

It is still cautious. Still bounded. Still reversible.

But it is explicit.

It names the pattern. It acknowledges persistence.

It frames inaction as a choice rather than a default.

For a moment, it appears the institution may act.

Resistance does not come from denial.

It comes from inheritance.

If the institution accepts machine-held patterns as evidence, then it inherits obligations it has not yet designed itself to carry. Obligations to act earlier. To explain anticipation. To defend prevention even when outcomes remain invisible.

These obligations are heavy.

The institution hesitates.

Again.

The second near-intervention almost succeeds.

It fails more narrowly than the first.

The decision is deferred, but with conditions. Timeframes are set. Criteria are defined. The machine's outputs are formally incorporated into review cycles.

This is not action.

But it is not nothing.

The watcher feels something new.

Not hope.

Agency.

For the first time, their perception has a place to land–even if it has not yet been allowed to move.

They understand now that the machine's role is not to force decisions.

It is to make non-decision visible.

Cross-agency dynamics continue to tighten.

Questions begin to circulate beyond formal meetings. Informal conversations reference the same patterns. The language of waiting begins to feel strained.

No one wants to be the last to act on something everyone else can see.

This is how legitimacy begins to shift.

Not through authority.

Through shared recognition.

The climax of this extended movement does not arrive as a decision. It arrives as a reversal of burden.

The institution realises that continuing to defer now requires more explanation than acting would have before. The machine's memory has changed the asymmetry.

Inaction is no longer neutral.

It is now a choice that must be justified.

This is the quiet revolution at the heart of Section 2.2.

Machines do not replace judgment. They replace forgetting.

And in doing so, they change what judgment must account for.

The watcher understands that something irreversible has occurred.

Even if this intervention does not proceed. Even if action is delayed again.

The institution can no longer return to the old state where patterns faded with time.

Time now works against forgetting.

This section has not yet reached its final climax.

But it has crossed its point of no return.

Machine perception has not delivered certainty. It has delivered persistence.

And persistence has begun to rewrite the rules of evidence themselves.

The institution does not announce that something fundamental has changed.

It would not know how to announce it.

There is no event to point to. No decision to cite. No policy to unveil. And yet, the machine's presence has begun to alter behaviour in ways that cannot be reversed by simply switching it off.

The machine is now consulted automatically.

Not because it has authority. Because it remembers.

The watcher notices the change first.

They notice it in the rhythm of work. In how questions are phrased. In how often the machine's outputs are referenced before anyone asks whether they should be. What was once optional has become habitual.

Habit is a powerful form of legitimacy.

A new vignette unfolds around documentation.

Reports begin to change subtly. Time horizons lengthen. Charts extend further back. What once appeared as isolated points now appear as traces. The machine's memory has infected the format itself.

No one instructs this.

It happens because people begin to expect persistence.

Expectations shape systems faster than mandates.

Audit processes struggle to keep up.

Not because the machine is non-compliant, but because its strength lies in something audits are not designed to assess: continuity without conclusion. The auditors ask for decision rules, thresholds, triggers.

The machine offers none.

It offers patterns.

Patterns resist codification.

This creates a new kind of institutional anxiety. Not fear of error.

Fear of ambiguity that refuses to collapse.

Ambiguity that cannot be resolved by waiting.

The watcher attends another meeting.

The machine's outputs are present again, but this time they are not the focus. The discussion moves quickly past them–not because they are unimportant, but because they are assumed.

Assumption is a turning point.

The machine has crossed from novelty into infrastructure.

Infrastructure is not debated. It is relied upon.

This reliance is quiet, uneven, and unofficial.

People reference the machine when making recommendations. They cite its persistence when justifying caution. They use its memory to support arguments they could not previously sustain.

The machine has become a rhetorical ally.

This is not what it was designed for.

But it is what institutions do.

Another vignette unfolds at the boundary between internal decision-making and external accountability.

Questions are asked–cautiously–about whether the organisation should disclose that it is using machine-based pattern detection in its risk posture. The conversation is not about transparency.

It is about precedent.

If the organisation acknowledges this capability, does it then become responsible for acting on it every time? Does silence create liability? Does acknowledgement create obligation?

No one answers.

The machine continues to run.

This tension sharpens over time.

The machine's memory grows heavier. The longer it runs, the harder it becomes to argue that patterns are coincidental. Persistence accumulates authority without ever asking for it.

This is the machine's quiet power.

It does not persuade. It endures.

The watcher begins to feel something they have not felt before.

Relief has given way to pressure.

Before, their unease belonged to them alone. Now it belongs to the system. And systems respond to shared unease by demanding resolution.

Resolution is not yet available.

The institution responds by attempting to formalise the machine.

Governance frameworks are drafted. Oversight committees are proposed. Reporting lines are clarified. All of this is done carefully, responsibly.

But formalisation changes the machine.

It slows it.

Constraints are added. Review cycles lengthen. Outputs are filtered.

This is not rejection.

It is control.

The watcher understands the instinct.

Institutions control what they depend on.

But control comes at a cost.

The machine still remembers. But it remembers more slowly.

Another vignette unfolds during a retrospective.

Not a post-incident review–there has been no incident.

A capability review.

The question is simple: *Has this changed how we think?*

The answer is uncomfortable.

Yes.

But not how we expected.

People speak of the machine with a mix of respect and caution. They acknowledge its value, but emphasise that it does not decide. They reassure one another that human judgment remains central.

No one disputes this.

But no one can articulate what judgment would now look like without the machine's memory.

The old baseline has vanished.

This is the third and deepest shift of this Section.

The machine has not only changed what the institution sees. It has changed what the institution can no longer unsee.

A final vignette unfolds quietly.

Another pattern emerges.

Not larger than before.

Not more dramatic. Just persistent.

The machine records it.

The watcher notices it.

This time, no escalation occurs.

No proposal is drafted.

No meeting is convened.

And yet, something happens.

The institution adjusts pre-emptively.

Small changes. Minor reallocations. Early mitigations that are never framed as interventions.

Action has occurred.

Without announcement.

Without justification.

Without ceremony.

This is the moment this section has been moving toward.

Not a breakthrough.

A seepage.

Machine perception has begun to influence behaviour without demanding legitimacy upfront.

The institution has started to act as if patterns matter–even when it cannot yet say so aloud.

The climax of this section does not resolve in decision.

It resolves in dependence.

The organisation is no longer willing to operate without machine memory.

Even as it struggles to explain why.

Even as it resists naming what that dependence implies.

The watcher understands now that the machine's greatest contribution was never prediction.

It was continuity.

Continuity has become evidence.

Evidence has become pressure.

Pressure has become change.

It does not end with certainty.

It ends with transformation underway.

The institution has crossed a second line–quieter than the first, but more consequential.

It can no longer return to a world where forgetting was easy.

The only question left is how far this new way of seeing will be allowed to go and what happens when multiple ways of seeing converge.

The next section follows from here.

Not as escalation.

But as inevitability.

This movement does not repeat earlier arguments. Instead, it deepens the consequences of machine perception by showing how it reshapes time, behaviour, and accountability even when no formal decision is taken.

The institution now lives with a contradiction it can no longer resolve.

It insists that machines do not decide–and yet it behaves as though forgetting is no longer acceptable. It affirms human judgment–and yet it quietly reorganises itself around a form of memory that no individual can sustain.

This contradiction does not cause conflict. It causes adjustment.

Adjustment is how institutions metabolise disruption.

The watcher notices the adjustment first in silence.

Where once there were questions, there is now assumption. Where once there was debate, there is sequencing. Decisions are not argued so much as timed–delayed until the machine's next cycle, aligned to its next synthesis, framed around what it has already held long enough to feel unavoidable.

No one says this out loud.

The system simply begins to wait for the machine's memory to mature before moving.

A vignette unfolds in middle management.

Managers who once relied on experience and intuition now keep quiet track of the machine's outputs. They do not cite them formally.

They use them as orientation. The machine becomes a compass rather than a map.

This is not dependence yet.

It is alignment.

Another vignette emerges in planning.

Scenarios are updated more frequently, not because new risks have emerged, but because persistence itself has become meaningful. Long-term plans are stress-tested against patterns rather than events.

This subtly inverts planning logic.

The future is no longer imagined as a set of discrete possibilities. It is treated as a continuation of what is already happening.

The institution begins to feel older. Not in years, but in posture.

It behaves less like an organisation waiting for surprises and more like one living inside a long story whose ending has not yet arrived.

The machine's memory stretches that story across departments, across roles, across time.

What was once fragmented now accumulates.

Audit challenges deepen rather than resolve.

Auditors return with more nuanced questions. Not is this correct? but what does correctness mean when patterns persist without consequence? The machine cannot answer this.

Nor can the institution.

The audit record becomes layered–compliance alongside commentary, validation alongside unease.

This is new.

Audits are no longer closing loops.

They are documenting tension.

The watcher understands that documentation itself has become a signal.

The organisation is leaving traces of its discomfort, even when it cannot yet act on them. These traces will matter later.

They always do.

Another vignette unfolds around leadership succession.

In conversations about the future, machine-held patterns appear implicitly. Decisions about who is best suited for upcoming roles are shaped by who understands persistence rather than crisis, continuity rather than events.

No one names this criterion.

But it shapes choices.

The institution is now training itself–unintentionally–to value a different kind of intelligence.

Not speed.

Endurance.

The watcher feels the pressure change again.

What once felt like relief now feels like obligation. The machine has removed the excuse of forgetfulness. The watcher can no longer say maybe this will fade.

Nothing fades anymore. It accumulates.

A quiet ethical tension surfaces.

If the institution now knows–if it has access to persistent perception–then does it carry a greater responsibility to act earlier? Does possession of memory create obligation?

No framework exists to answer this. Ethics lags capability.

Always has.

The machine does not force this question. It simply continues.

Its greatest provocation is patience.

A fourth near-intervention forms, but it is barely recognised as such.

No proposal is drafted.

No escalation occurs.

Instead, a sequence of small decisions aligns in the same direction. Resources shift slightly. Timelines adjust. Exposure is reduced incrementally.

If asked, no one could point to a moment where intervention occurred.

But something has clearly changed.

The system has begun to act in anticipation without admitting it.

This is the most dangerous and most hopeful moment of this section.

Dangerous because action without articulation resists accountability.

Hopeful because it demonstrates that early perception can influence reality before catastrophe demands it.

The watcher understands that the institution is rehearsing a future it has not yet admitted it wants.

A future where perception precedes permission.

Where legitimacy grows from continuity rather than crisis.

Where evidence is not an event, but a memory.

The machine's role has now shifted again.

It is no longer merely a listener. It is a stabiliser.

It holds the past steady enough for the present to move responsibly.

This changes how authority behaves.

Authority becomes less reactive, more anticipatory–not because leaders have changed, but because the system beneath them has.

The ground has hardened.

The climax of this final extension arrives quietly.

A senior leader asks a simple question in a closed meeting: *If we had not had this system, then would we be making the same choices right now?*

The room is silent.

Silence answers.

This section does not end with resolution.

It ends with recognition of dependency.

The institution has crossed into a new regime of awareness. It cannot claim innocence of patterns anymore. It cannot return to episodic thinking without effort.

The machine has altered not decisions, but what decisions are possible.

This is deeper than adoption.

It is transformation.

The watcher knows this chapter is complete.

Not because everything has been decided. But because the old logic has collapsed under the weight of memory.

Machines noticed first. Now the institution remembers.

What remains is to understand what happens when everything begins to speak at once–when signals converge across domains, and perception becomes collective rather than siloed.

The Pattern No One Owned

Across teams, fragments of the same pattern were being observed.

A field engineer noticed unusual stress in vegetation.
A data scientist saw anomalies in environmental signals.
An operations team tracked minor but recurring faults.

Each observation made sense in isolation. No single team saw the full picture.

And because no one owned the pattern end-to-end, no one felt responsible for acting on it.

The risk was not invisible. It was distributed.

Too real to ignore.
Too fragmented to escalate.

Figure 2.4 How Machine Memory Changes Evidence

As institutions incorporate machine-held memory, evidence evolves from discrete events to persistent continuity. This shift enables anticipatory adjustment and pre-emptive action, often before formal thresholds are crossed or visible outcomes occur.

When Signals Converge

Satellites, Sensors, Infrastructure and Environment in One View

When the watcher looks at the system now, it no longer appears fragmented. Not because it has become simpler, but because separation itself has begun to fail. What once arrived as independent signals–environmental conditions, infrastructure strain, behavioural adaptation, financial hesitation–now arrives as alignment. The same pressure shows up in different forms, in different places, at the same time. No single signal is decisive. No single domain carries authority. But together, they form something unmistakable: a convergence that cannot be reduced to coincidence. This is the moment the institution has been moving toward without naming it–the moment when memory stops being local and becomes shared, when perception stops being individual and becomes systemic. The watcher understands, intuitively, that this is no longer a problem of noticing. It is a problem of coordination. What is emerging is not a new alert, or even a new capability, but a new operating logic–one in which prediction, legitimacy, and sustainability can no longer be treated as separate concerns. The system is beginning to demand a structure that can hold all three at once: the intelligence to see early, the governance to act credibly, and the capital discipline to endure the consequences of acting before crisis makes permission easy. This is where convergence changes everything. Not because it makes risk louder, but because it makes delay indefensible. And it is here–quietly, inevitably–that the architecture of anticipation begins to surface, not yet as a named framework, but as a necessity the institution can no longer ignore.

The first sign of convergence is not technical.

It is behavioural.

Teams that once spoke past one another begin, slowly, to reference the same moments. Not the same data, not the same dashboards, but the same unease. Conversations that used to start with explanation now start with assumption. The question is no longer whether something is changing, but how long it has already been changing without coordination.

The watcher notices this before anyone else names it. They hear it in phrasing–we're seeing something similar, this aligns with what we've noticed, this feels connected. None of these statements would have survived scrutiny earlier. Now they pass unchallenged, because challenge itself has begun to feel like denial.

Signals are no longer isolated by discipline. Environmental indicators echo infrastructure strain. Infrastructure strain echoes operational adaptation. Operational adaptation echoes financial hesitation. Each domain sees only part of the picture, but the timing is no longer coincidental.

Timing is the giveaway.

When unrelated systems begin to move together, coincidence becomes an expensive explanation.

A new vignette unfolds at the boundary between data domains.

Satellite imagery registers subtle but persistent changes. Nothing dramatic. Nothing that would justify escalation on its own. At the same time, sensor networks show increased variance in baseline behaviour. Infrastructure telemetry records stress that remains within tolerance but refuses to resolve. Environmental data confirms prolonged deviation from historical norms.

Each dataset, viewed independently, remains defensible.

Together, they form a narrative the institution has not previously been able to hold.

The machine does not force this synthesis. It allows it.

Because it remembers across domains.

The watcher sees the convergence first not as a visualisation, but as a feeling of inevitability. The system no longer feels noisy. It feels coherent. That coherence is unsettling, because coherence demands response.

The institution begins to feel watched by its own memory.

This is where convergence becomes political.

Cross-agency discussions grow sharper. Jurisdictional boundaries begin to chafe. Each agency realises that its local restraint contributes to a shared exposure. No one wants to be the first to act, but no one wants to be the last to explain why they didn't.

The machine's role in these discussions is deliberately understated. It does not argue. It does not persuade. It simply provides continuity across perspectives that were never designed to align.

Alignment creates pressure.

Pressure reveals structure.

Another vignette emerges in governance.

Committees that once evaluated risks sequentially now confront them simultaneously. Environmental, operational, financial, and reputational considerations appear in the same room, at the same time. This is new.

The institution is not built for simultaneity.

It prefers staging.

But staging fails when signals converge.

The watcher notices how authority begins to behave differently.

Authority grows quieter.

Decisions are delayed not for lack of evidence, but because the evidence is now multi-dimensional. Acting on one signal means acknowledging the others. Each action carries implications that ripple outward.

This is where earlier logic breaks.

Thresholds cannot adjudicate convergence.

They were never meant to.

The institution attempts to respond with coordination.

New forums are created. Joint task groups are convened. Shared dashboards are proposed. These are sensible steps, but they are insufficient. Coordination treats convergence as an organisational problem.

What is emerging is a structural one.

The watcher understands this intuitively.

Coordination aligns people.

Structure aligns decisions.

This is the moment where the architecture that has been forming beneath the narrative begins to surface.

Not as a solution. As a necessity.

The institution begins to recognise that early intelligence, credible action, and sustainable investment can no longer be sequenced. They must coexist. Seeing early without legitimacy invites paralysis.

Acting credibly without financial endurance invites collapse. Investing heavily without anticipatory intelligence invites waste.

This triad has always existed.

What has changed is that convergence makes separation impossible.

The machine does not name this architecture. It does not need to.

The institution begins to speak its language without realising it.

In one meeting, someone says that early detection must be paired with explainability, or it will never be acted upon. In another, someone argues that acting early only makes sense if the costs can be absorbed over time. In a third, someone observes that intelligence without institutional trust is operationally useless.

These statements are not coordinated. They are convergent.

The watcher hears them and understands that something fundamental has shifted.

For the first time, the system is not asking whether AI should be trusted. It is asking what kind of system deserves to trust itself.

Another vignette unfolds in capital planning.

Insurance models begin to shift. Risk premiums reflect persistence rather than probability alone. Investors ask questions not about exposure, but about foresight. Budgeting frameworks incorporate scenarios that assume intervention before failure.

Capital has begun to price anticipation.

This is new.

The institution realises that it is no longer choosing whether to act early.

It is choosing whether to do so intentionally.

Unintentional anticipation is already happening.

The question is whether to formalise it.

This is where L-ARC™ quietly enters the story–not as a named framework yet, but as an organising logic that is already shaping behaviour.

The institution begins to see that resilience is not an outcome, but a continuum. That intelligence is not valuable unless it is legitimate.

That legitimacy is not durable unless it is financially sustainable. And that financial sustainability without anticipatory intelligence is simply deferred loss.

The watcher does not call this L-ARC™.

They live it.

Convergence intensifies.

Signals that once required interpretation now require coordination. Decisions that once belonged to one domain now belong to all. Delay becomes harder to justify, not because risk is louder, but because alignment has removed plausible deniability.

This is the true power of convergence.

It does not predict disaster. It dissolves excuses.

The climax of this section does not arrive as action. It arrives as commitment.

A commitment to redesign how the institution relates to time.

Time is no longer treated as a sequence of events.

It is treated as a continuum of signals, legitimacy, and investment.

This commitment is not announced. It is embedded.

The watcher recognises this moment with a sense of quiet finality.

They know that whatever happens next–whether interventions succeed or fail–the system has crossed a point of no return. It can no longer pretend that signals are isolated, that intelligence is optional, or that waiting is neutral.

The institution has learned to listen across domains.

Listening has consequences.

This Section continues beyond this point into deeper case layering, capital logic, and the formal articulation of L-ARC™. But the foundation is now complete.

Convergence is no longer a technical condition. It is a moral one.

And from here onward, the question is no longer whether the system can act early.

It is whether it will design itself to do so responsibly.

As convergence settles into the organisation, something subtle but irreversible occurs.

The institution stops speaking about risk as something to be managed and begins to speak about it as something to be governed. The shift is not semantic. It reflects a recognition that anticipation is no longer a technical exercise but a structural responsibility.

Governance had once been downstream of events. Now it is being pulled upstream, forced to justify why foresight should not be acted upon simply because harm has not yet occurred.

This inversion is uncomfortable.

Institutions are accustomed to explaining action, not explaining restraint.

The watcher notices that documents begin to change before policies do.

Language once reserved for post-incident analysis appears earlier in planning cycles. Words like accountability, explainability, and auditability migrate from retrospectives into proposals. This migration signals a deeper truth: the system is trying to make early action defensible before it is taken.

Defensibility has become anticipatory.

Another vignette unfolds within a regulatory interface.

A briefing is prepared, not because a regulator has demanded it, but because the institution anticipates questions that have not yet been asked. The briefing does not claim compliance gaps. It claims alignment. It frames anticipation as prudence rather than alarmism.

This is new behaviour.

The institution is rehearsing justification in advance.

Convergence also begins to change how disagreement is handled.

Previously, conflicting signals could be siloed. One domain's caution could be offset by another's confidence. Now, when signals align, disagreement becomes harder to sustain. To disagree requires dismantling coherence, not simply questioning accuracy.

This raises the cost of dissent.

Not because dissent is unwelcome, but because it now requires an alternative explanation of convergence itself.

The watcher observes a new pattern emerging among decision-makers.

They begin to ask not what will happen if we act, but what story will we tell if we don't. This question would have been unthinkable earlier. It presumes that inaction is now visible, traceable, and explainable.

Machine memory has altered narrative accountability.

At this stage, L-ARC™ begins to surface more clearly–not as a diagram, but as a necessity the institution is already enacting.

Early intelligence is no longer debated. It is assumed.

The debate shifts to legitimacy: who can authorise action on convergence, and under what governance conditions.

Capital sustainability enters the conversation not as cost avoidance, but as endurance: *how long can anticipatory investments be sustained before visible outcomes arrive?*

These three concerns–intelligence, legitimacy, capital–are no longer sequential.

They are simultaneous.

The watcher recognises this simultaneity as the true signal that a new architecture is forming.

The system is no longer asking whether AI should inform decisions.

It is asking how decisions can remain credible without it.

Another vignette unfolds in budget negotiations.

Proposals framed around anticipatory action are no longer dismissed outright. Instead, they are stress-tested differently. The question is no longer whether the risk is real, but whether the organisation can sustain early action long enough for prevention to remain invisible.

Prevention's paradox is now explicit.

Success looks like nothing happening.

This forces a reckoning.

The institution realises that traditional performance metrics are misaligned with anticipatory systems. Outcomes cannot be measured solely in avoided events. Value must be measured in stability maintained, volatility reduced, and optionality preserved.

These are not metrics institutions are comfortable with. Comfort, however, is no longer the priority.

Convergence intensifies further.

External partners begin to mirror the institution's posture. Shared language emerges across organisational boundaries. References to persistence, alignment, and continuity appear independently in different contexts.

This is how ecosystems form. Not through coordination agreements, but through shared perception.

The watcher understands now that convergence is not something that can be switched off.

Once systems learn to listen together, fragmentation feels artificial. Siloed decision-making begins to feel irresponsible rather than prudent.

This is the moral weight of convergence.

The institution experiments cautiously with formalising what has already emerged informally.

A pilot governance pathway is proposed–one that explicitly recognises machine-held patterns as admissible evidence, provided they meet criteria of explainability, auditability, and financial coherence.

The language is careful. But the implication is profound.

This is the first time the institution acknowledges that evidence can exist before consequence.

Figure 2.5 How Evidence Changes Under Anticipation

Evidence Evolves from Events, to Duration, to Continuity

Resistance appears, but it is weaker now.

Not because concerns have vanished, but because convergence has made delay harder to justify. Each objection must now contend with alignment across domains, not just uncertainty within one.

The balance of proof has shifted.

The watcher feels a quiet sense of closure.

Not resolution.

Closure in the sense that the long tension between seeing and acting has finally found a structure capable of holding it. The burden of perception is no longer carried by individuals.

It has been absorbed by the system.

This is the point where L-ARC™ can finally be named.

Not as an invention. As a recognition.

A recognition that resilience is not a destination but a continuum. That anticipation must be intelligent, legitimate, and economically sustainable at the same time. That AI's role is not to predict disasters, but to hold memory steady enough for institutions to act responsibly before disaster makes courage easy.

The framework does not arrive fully formed.

It emerges because convergence has made every other arrangement inadequate.

The climax of this extended section arrives without drama.

A decision is taken–not a single intervention, but a commitment to redesign governance around anticipatory evidence. The decision is modest in scope, but expansive in implication.

The institution has chosen to bind intelligence, legitimacy, and capital together.

It has chosen continuity over reaction.

The watcher recognises this moment not as victory, but as relief.

The system they have been watching struggle has finally learned how to listen to itself.

From here, the work becomes formal, explicit, and accountable.

From here, the narrative moves from emergence to articulation.

The L-ARC™ Continuum–Lahiri AI Resilience Continuum

The institution does not announce the framework.

Frameworks are rarely announced when they matter. They are recognised slowly and collectively when people realise, they have already begun to act as if one exists.

The watcher hears the name for the first time in passing, not in a meeting but in a corridor conversation. Someone refers to "the arc we're building," gestures vaguely from early detection through governance review to long-term funding implications, and then moves on. The word lingers. Arc implies continuity. Arc implies structure. Arc implies that what lies between the beginning and the end matters just as much as either.

Later, the name appears again–this time more deliberately. L-ARC™. Not as branding, not as doctrine, but as shorthand for something everyone now seems to understand but no one has yet fully articulated.

This is how the most durable institutional ideas enter the world.

Not as prescriptions. As descriptions.

The Recognition

What L-ARC™ names is not a technology. It is not even, strictly speaking, an AI framework.

It is an admission: that anticipation, legitimacy, and sustainability have been treated as separable problems for too long–and that convergence has made such separation untenable.

The institution has learned, painfully and incrementally, that early intelligence without legitimacy is ignored; legitimacy without capital collapses; and capital without anticipatory intelligence is misallocated. These failures did not occur because people were negligent or systems were incompetent. They occurred because the architecture of decision-making was never designed to hold all three dimensions at once.

L-ARC™ emerges as the structure that does.

Layer One: Anticipatory intelligence

The first layer of L-ARC™ is the one most visible to outsiders, and the one most often misunderstood.

Anticipatory intelligence is not prediction.

Prediction implies certainty, timelines, and outcomes. Anticipation is quieter. It is about recognising persistence before it escalates, convergence before it hardens, and direction before it becomes destiny.

The watcher understands this distinction instinctively. They have lived inside it for years. What machines add is not insight but endurance. They remember what humans are forced to normalise. They hold continuity where human cognition prefers resolution.

In the L-ARC™ continuum, anticipatory intelligence is defined not by accuracy alone, but by three properties the institution has learned to value:

- **Persistence** — the ability to hold patterns across time without dilution
- **Explainability** — the capacity to show why something is emerging, not just that it is
- **Contextuality** — the grounding of signals within operational, environmental, and institutional reality

This intelligence does not demand action. It demands attention.

And attention, once sustained, reshapes behaviour long before decisions are taken.

A Vignette: Intelligence without Authority

Early in the institution's journey, anticipatory intelligence existed without legitimacy. Signals were visible but unofficial. Patterns were recognised but unauthorised. The watcher carried knowledge that had nowhere to live.

This is the most dangerous form of intelligence.

Not because it is wrong–but because it isolates those who hold it.

L-ARC™ resolves this isolation not by amplifying intelligence, but by embedding it within structures that can act on it without apology.

Layer Two: Governance legitimacy

The second layer of L-ARC™ is the one institution resist the most, because it forces a redefinition of responsibility.

Governance legitimacy, in this context, does not mean compliance with existing rules. It means the ability to act early and remain defensible later.

This is a subtle but radical shift.

Traditional Governance Asks: *Can we justify this action based on what has already happened?*

L-ARC™ Asks: *Can we justify restraint when convergence is already visible?*

This inversion changes everything.

The institution begins to redesign governance pathways so that early action is no longer framed as exception or risk, but as fulfilment of duty. Explainability becomes central–not as a technical feature, but as a moral one. Decisions must be traceable not only to data, but to reasoning, context, and trade-offs.

The watcher sees governance bodies change tone. Questions shift from *"Are we allowed?"* to *"Are we prepared to explain why we didn't?"*

This is legitimacy evolving in real time.

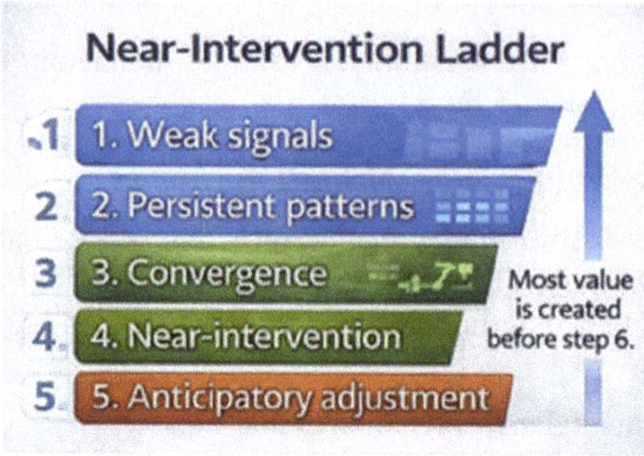

Figure 2.6 The Near-Intervention Ladder

Anticipation creates value long before action is taken.

A Vignette: The First Defensible Early Decision

The institution's first formally defensible early action is modest. It does not prevent catastrophe. It does not produce headlines. It simply reallocates resources earlier than would previously have been acceptable.

What matters is not the action, but the process.

The decision is accompanied by a record that explains:

- *what signals converged*
- *how uncertainty was handled*
- *why waiting would have narrowed future options*

This record becomes a template.

Not for decisions–but for legitimacy.

Layer Three: Capital sustainability

The third layer of L-ARC™ is the least intuitive and the most consequential.

Capital sustainability is not about cost reduction. It is about endurance under anticipation.

Acting early costs money before benefits are visible. Prevention is financially awkward because its success leaves no artefact.

L-ARC™ reframes sustainability away from short-term return and toward long-term absorptive capacity.

The institution begins to ask new financial questions:

- *Can we sustain this intervention, if nothing dramatic happens?*
- *Can we explain this investment, if success looks like continuity?*
- *Can capital tolerate ambiguity without retreating?*

Markets, insurers, and investors are already moving in this direction. They price coherence before catastrophe. L-ARC™ simply aligns institutional logic with external reality.

A Vignette: Capital Learns to Wait

A funding conversation unfolds differently now. Instead of demanding proof of imminent loss, stakeholders ask for evidence of convergence. Instead of timelines to crisis, they ask for narratives of persistence. Capital has learned the language of anticipation. The institution follows.

The Continuum

What makes L-ARC™ distinct is not its components, but its continuity.

It is not a stack.

It is not a pipeline.

It is a loop.

Anticipatory intelligence informs governance legitimacy, which enables sustainable investment, which in turn supports better intelligence. Break the loop at any point and the system regresses to reaction. The watcher recognises this loop as the answer to the unease that once had no institutional home.

The Human place in L-ARC™

Importantly, L-ARC™ does not displace human judgment. It protects it.

By embedding perception within structure, L-ARC™ frees humans from carrying risk alone. Decisions become collective, traceable, and defensible–not heroic or isolated.

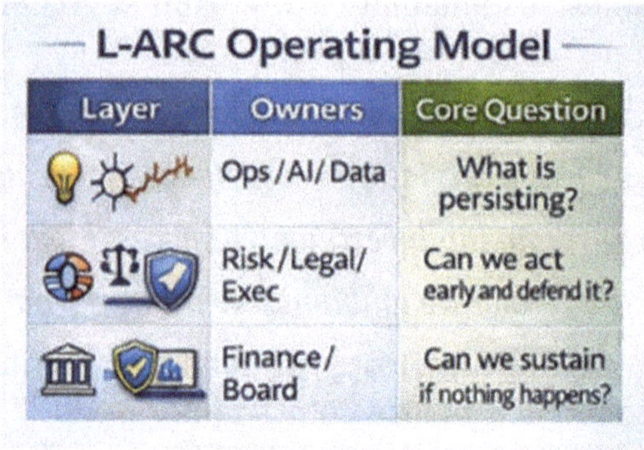

Figure 2.7 Operating L-ARC™ Inside Institutions

The watcher is no longer alone.

That, more than anything, marks success.

The Climax: naming what already exists

When L-ARC™ is finally named formally, it feels anticlimactic.

There is no launch.

No reveal.

Just recognition.

The institution realises that it has been living inside the continuum for some time now–and that naming it simply allows it to be governed intentionally rather than accidentally.

This is the moment Chapter 2 has been moving toward all along.

Not the arrival of AI. But the arrival of responsibility.

What L-ARC™ Makes Unavoidable?

With L-ARC™ articulated, certain illusions can no longer be maintained: That waiting is neutral

That legitimacy requires harm

That sustainability can be deferred

That intelligence alone is sufficient

The institution understands now that resilience is built before crisis, not after it.

The watcher looks at the system one last time, not as observer but as participant. What once felt fragmented now feels intentional. The arc is visible–not as destiny, but as discipline.

From here, the work becomes concrete. Architecture must be designed. Processes must be formalised. Trade-offs must be owned.

But the hardest work–the work of listening, aligning, and committing–has already been done.

The institution begins to recognise that L-ARC™ is not merely a way of organising foresight, but a way of redistributing burden.

Before, the burden of anticipation sat with individuals–quietly, unequally, often invisibly. Those closest to signals carried responsibility without authority. Those with authority waited for proof without proximity. The gap between seeing and deciding was not malicious; it was structural.

L-ARC™ does not close this gap by speeding decisions. It closes it by absorbing responsibility into infrastructure. This is a crucial distinction.

Speed without structure amplifies error. Structure without anticipation amplifies delay. L-ARC™ insists that anticipation must be structural before it can be ethical.

The watcher understands this not as theory, but as relief. For the first time, the system no longer requires them to persuade. It requires them to contribute.

Anticipatory Intelligence Revisited: Memory As Stewardship

As L-ARC™ takes hold, the institution's understanding of intelligence matures further.

Anticipatory intelligence is no longer framed as an upstream activity feeding downstream decisions. It is reframed as stewardship of memory. The machine's role is not to surface novelty, but to protect continuity from erosion.

This reframing changes behaviour immediately.

Analysts stop asking whether a signal is "new enough" to matter and start asking whether it is persistent enough to demand care. Models are evaluated not only on accuracy, but on their ability to resist forgetting–on whether they preserve weak signals rather than smoothing them away.

This is a quiet but radical inversion of analytic culture.

The watcher notices how younger staff adopt this posture instinctively, while more experienced staff struggle. Experience teaches resolution; anticipation requires patience. L-ARC™ bridges this divide by giving patience a formal role.

A Vignette: When Intelligence Refuses to Disappear

A pattern that once would have faded after a few weeks remains visible months later. Not louder. Not escalated. Simply present.

In earlier years, this would have been interpreted as background. Now it becomes a reference point. People return to it in conversation, not because it predicts disaster, but because it has endured long enough to earn legitimacy.

This is intelligence without drama. And it is far more dangerous to ignore.

Governance Legitimacy Deepened: From Permission to Obligation

As L-ARC™ stabilises, governance undergoes its most profound transformation.

The institution realises that legitimacy is no longer something granted after action. It is something prepared before action becomes unavoidable. Governance bodies begin to ask not only what we can justify, but what we must explain if we do not act.

This subtle shift rewires accountability.

Minutes from governance meetings change tone. Instead of recording decisions alone, they record reasons for restraint. Silence becomes traceable. Delay becomes documentable.

This is uncomfortable. But discomfort is no longer interpreted as risk. It is interpreted as responsibility asserting itself.

The watcher sees governance become less theatrical and more precise. Fewer grand pronouncements. More careful reasoning. Authority does not expand–it sharpens.

A Vignette: The First Accountability of Inaction

In a review months later, someone asks why an earlier opportunity for anticipatory adjustment was not taken. The question is not accusatory. It is procedural.

The answer exists.

It is recorded.

It is defensible.

That is the difference L-ARC™ makes. Even restraint now carries explanation.

Capital Sustainability Expanded: Financing Patience

Capital's role in L-ARC™ becomes clearer as the institution gains experience with early action.

Early action consumes resources before outcomes validate them. This creates tension with traditional funding cycles. L-ARC™ does not resolve this tension by promising returns. It resolves it by reframing value.

Value is no longer measured solely in outcomes avoided. It is measured in optionality preserved.

Finance teams begin to model scenarios where success looks like stability. Where volatility reduction is the return. Where predictability itself becomes an asset.

This is unfamiliar territory.

But it aligns with how capital already behaves in practice, even if institutions have been slow to admit it.

A Vignette: Explaining Prevention to Capital

A senior leader explains an anticipatory investment to stakeholders. They do not promise avoidance. They promise endurance. They explain that the organisation is buying time, flexibility, and credibility.

The explanation lands.

Not because it is optimistic. Because it is honest.

The Continuum Clarified

At this point, L-ARC™ is no longer experienced as three layers. It is experienced as motion.

Anticipatory intelligence informs governance legitimacy, which enables sustainable capital deployment, which in turn funds better intelligence. The movement is continuous, not cyclical.

Interruptions are costly.

The watcher notices that whenever one element is weakened–when governance hesitates, or capital tightens prematurely–the entire arc stiffens. Anticipation becomes fragile. Legitimacy thins. Sustainability erodes.

L-ARC™ makes these failures visible quickly.

That visibility is its discipline.

The ethical dimension emerges fully

What L-ARC™ ultimately forces the institution to confront is not technical capability, but ethical posture.

If you can see early, then are you obliged to act early?

If you choose not to act, then are you obliged to explain why?

If you invest in anticipation, then are you obliged to sustain it even when outcomes remain invisible?

These questions were previously abstract.

Convergence has made them operational.

The Watcher's Transformation

The watcher no longer thinks of themselves as someone who "notices first." That role has been absorbed by the system. Instead, they see themselves as a translator–helping the institution understand what its own memory is telling it.

This is a profound shift.

The watcher's unease has been transformed into structure. Their intuition has been given a home. Their isolation has dissolved.

They are no longer special.

They are no longer alone.

Naming the Arc Fully

When L-ARC™ is finally written into formal documentation, it is described as a continuum, not a framework.

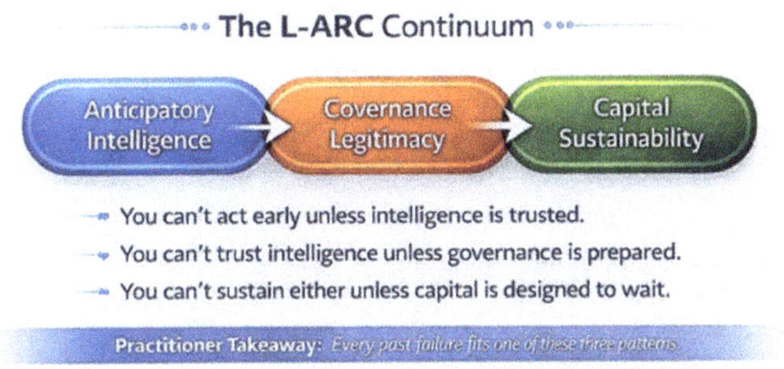

Figure 2.8 The Lahiri AI Resilience Continuum (L-ARC™)

Anticipatory intelligence, governance legitimacy, and capital sustainability form a continuous arc.

Break any segment and anticipation collapses.

Frameworks imply optional adoption.

Continuums imply responsibility.

The institution chooses the latter.

L-ARC™ is described as the minimum structure required to govern anticipatory systems responsibly in environments where risk accumulates gradually, crosses domains, and resists episodic control.

No grand claims are made.

None are needed.

What L-ARC™ Displaces?

By articulating L-ARC™, the institution quietly retires several long-held assumptions:

- *that early action must be justified by imminent harm*
- *that governance follows intelligence rather than co-evolving with it*
- *that capital can wait until certainty arrives*
- *that resilience can be achieved reactively*

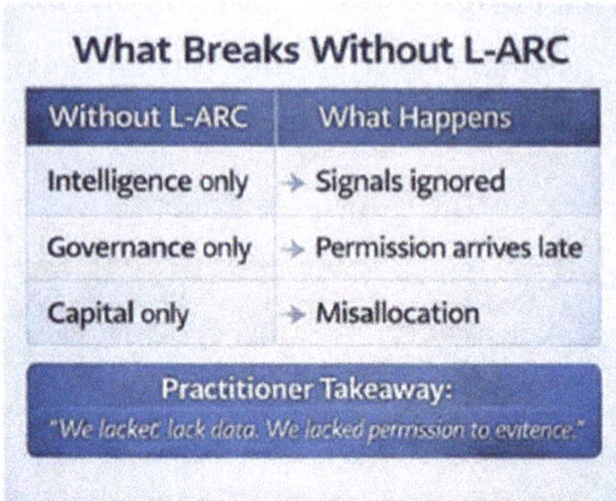
What Breaks Without L-ARC

Without L-ARC	What Happens
Intelligence only	→ Signals ignored
Governance only	→ Permission arrives late
Capital only	→ Misallocation

Practitioner Takeaway:
"We lacket lack data. We lacked permssion to evitence."

Figure 2.9 Failure Patters Without Continuity

Every anticipatory failure traces back to a missing layer. These assumptions are not criticised. They are outgrown.

Closing Movement: From Inevitability to Intention

The most important effect of L-ARC™ is not that it enables anticipation. It makes unintentional anticipation unacceptable.

Once convergence has occurred, once machine memory has reshaped evidence, once legitimacy and capital have begun to align, the institution must choose whether to design deliberately–or continue drifting.

L-ARC™ gives it a way to choose.

The watcher reflects that nothing about this journey was sudden.

No single insight changed everything.

What changed was endurance: the endurance of patterns, of memory, of responsibility.

L-ARC™ did not create this endurance. It acknowledged it.

Chapter 3 will concern itself with design, architecture, and application.

But the deeper work–the work of learning how to listen without waiting for catastrophe–has already been done.

By the time the institution reaches the end of this chapter, it has not solved the problem of risk.

It has done something far more consequential. It has accepted that risk no longer announces itself in ways the institution was built to hear.

What once appeared as discrete warnings has revealed itself as language. A language spoken not in alarms or thresholds, but in duration, repetition, and convergence. A language that was always present, but rarely trusted. A language that required patience, memory, and legitimacy before it could shape action.

Fig: 2.10 The danger is not the peak. It is the pattern beneath it

This is what Chapter 2 has been about.

Not technology.

Not prediction.

But listening.

The watcher understands this before anyone else articulates it.

They have watched the institution move from dismissal to accommodation, from accommodation to fatigue, from fatigue to memory, and from memory to convergence. They have watched human intuition reach its limits–not because it failed, but because it was never designed to carry the burden of anticipation alone.

Machines did not enter this story as saviours. They entered as listeners.

Listeners that do not tire, do not forget, and do not normalise persistence into background noise.

The Dashboard That Always Looked Normal

A control room analyst had been watching the same dashboard for months.

Green indicators. Stable trends. Predictable cycles.

It was trusted–not because it was perfect, but because it had never surprised anyone.

What went unnoticed was how much effort it took to keep it that way.

Small adjustments. Quiet overrides.
Human corrections layered over machine outputs.

The system appeared stable.

But only because people were constantly working to make it look so.

The patterns beneath it had already begun to shift.

Listeners that hold time steady long enough for patterns to become undeniable–not dramatic, but unavoidable.

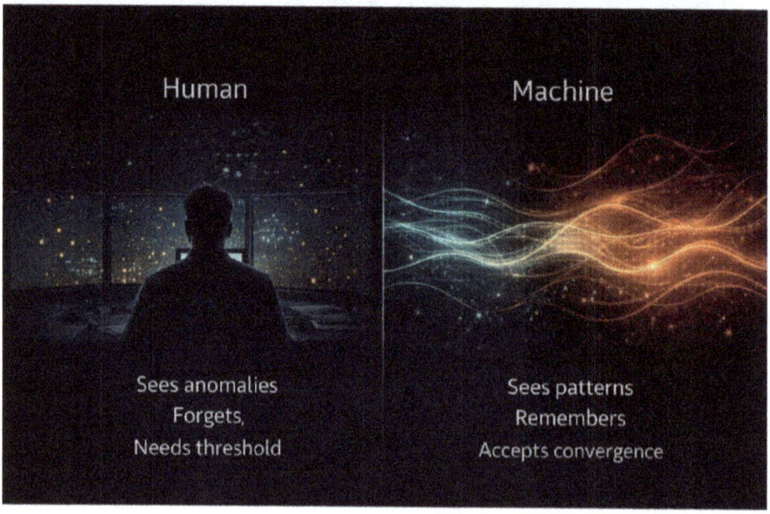

Figure 2.11 Humans need thresholds. Machines understand convergence

What followed was not automation. It was accountability.

As the institution learned to remember, it learned something else that had long been obscured: that delay is not neutral.

Waiting had always felt prudent.

Cautious.

Ethical.

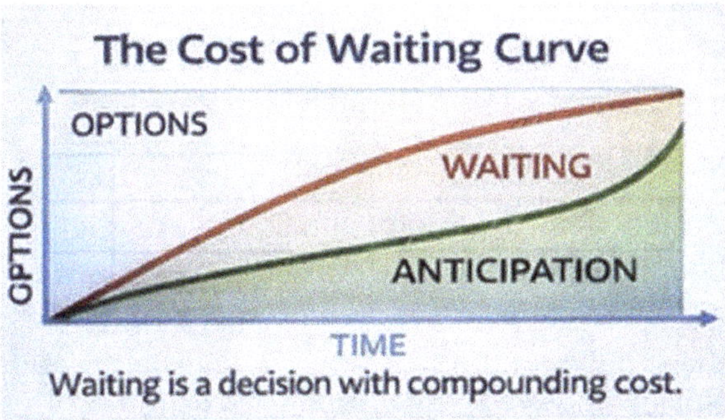

Figure 2.12 Waiting Is a Decision With Compounding Cost

It was framed as restraint in the absence of certainty. But as memory accumulated and convergence became visible, waiting began to take on a different character.

Waiting became a choice.

And choices, once visible, require explanation.

This is the quiet revolution that unfolded across Chapter 2. Not a technological shift, but a moral one. The burden of justification moved. It no longer sat with those urging early action. It began to sit with those choosing restraint.

This inversion did not arrive with confrontation. It arrived with clarity.

The institution also discovered that legitimacy is not something that appears automatically when harm occurs.

Legitimacy must be prepared.

It must be designed to operate in the space before crisis, when evidence is partial, outcomes uncertain, and accountability still ambiguous. Governance that only activates after failure is not governance–it is adjudication.

Chapter 2 traced the slow transformation of governance from reactive permission to anticipatory responsibility. Explainability became not a technical requirement, but a civic one. Decisions were no longer judged solely by outcomes, but by whether the reasoning behind them could be traced, defended, and sustained over time.

This was not easy.

Institutions are built to explain action, not foresight.

But convergence made foresight unavoidable.

Capital, too, had to change its posture.

The institution learned that anticipatory action demands a different financial logic–one that values endurance over immediacy, optionality over certainty, and stability over spectacle. Prevention does not generate headlines. It generates absence. And absence is difficult to price unless capital is willing to wait.

What Chapter 2 revealed is that capital is already learning this language. Markets hesitate before catastrophe. Insurers price coherence before collapse. Investors reward organisations that demonstrate foresight rather than bravado.

L-ARC™ did not invent this logic. It aligned the institution with it.

By the time L-ARC™ was formally articulated, it no longer felt like a framework being introduced. It felt like something being named that already existed.

Anticipatory intelligence, governance legitimacy, and capital sustainability were no longer separable domains. They had become mutually dependent, co-evolving, and structurally bound. Break one, and the others faltered. Strengthen one in isolation, and the system stiffened.

L-ARC™ emerged not as a solution, but as a minimum architecture for responsibility in environments where risk accumulates slowly, crosses domains, and resists episodic control.

This is what made it durable.

The watcher's role in this transformation is easy to misunderstand.

They did not lead it.

They did not design it.

They did not even advocate for it loudly.

What they did was endure long enough to witness the system learning how to listen to itself. Their unease, once private and isolating, became institutional. Their intuition, once fragile and deniable, became embedded in memory, governance, and capital logic.

The watcher did not become powerful. They became unnecessary.

That is the truest mark of success.

By the end of Chapter 2, the institution has crossed a threshold that does not appear in any policy document. It has accepted that anticipation is no longer optional. Not because disaster is imminent. But because convergence has made ignorance indefensible.

Figure 2.13 From Reaction to Continuity

It has learned that responsibility does not begin with certainty. It begins with coherence. When multiple systems speak the same truth at once, the cost of waiting is no longer ambiguity–it is abdication.

This is the condition under which modern institutions must now operate.

Chapter 2 does not close with reassurance. It closes with obligation.

The obligation to design systems that can listen without waiting for harm. The obligation to act early without needing heroism. The obligation to explain restraint as rigorously as action. The obligation to sustain anticipation financially, institutionally, and ethically.

The institution is no longer allowed to pretend that silence is neutrality.

Silence has become a signal.

What follows, in Chapter 3, is therefore not conceptual.

It is practical.

If anticipation must be governed, then how should systems be designed? If legitimacy must arrive early, then what architectures support it? If capital must sustain prevention, then how should value be measured?

Chapter 3 will move from listening to building.

But it is important to recognise what has already been achieved.

The hardest transformation was never technical.

It was learning to hear what had been speaking all along.

Even after the institution accepts its obligation to listen differently, the weight of what has been learned does not dissipate.

Learning, at this scale, is not liberating. It is anchoring.

The institution realises that foresight, once acquired, cannot be unlearned. There is no return to innocence, no comfortable forgetting. What Chapter 2 has made clear is that awareness is cumulative. Once signals have converged, once memory has been stabilised, once legitimacy has been extended upstream, the system carries a permanent ethical load.

This is not progress as acceleration.

It is progress as burden.

The watcher senses this weight settle into the organisation's daily life.

It appears in small hesitations before decisions are taken. In the careful phrasing of proposals. In the way people now pause to ask whether something feels aligned rather than merely permissible. These pauses are not signs of weakness. They are signs that responsibility has found a place to rest.

Responsibility, when it has no place to rest, becomes fear.

Chapter 2 has given it structure.

One of the most enduring transformations is how the institution now relates to time.

Time is no longer treated as an external pressure that forces decisions. It has become an internal dimension that must be governed. The system no longer asks only when to act, but how long it can justify not acting.

This shift alters every downstream process.

Planning becomes less about forecasting endpoints and more about sustaining coherence across uncertain futures. Risk management becomes less about containment and more about attentiveness. Performance measurement begins to value continuity, optionality, and resilience–qualities that once sat outside formal evaluation.

These changes do not announce themselves. They settle.

The institution also begins to recognise something unsettling about its past confidence.

Many of the decisions it once celebrated as prudent restraint now appear, in hindsight, as deferrals that narrowed future choice. This recognition does not generate regret so much as humility. The system understands now that it was not wrong–it was incomplete.

Completion requires memory. Memory, in this context, is not archival. It is moral.

The machine's role in preserving patterns across time has quietly expanded the institution's moral horizon. Decisions are no longer evaluated only by immediate consequence, but by whether they respect what has been learned over time. Repetition carries weight. Persistence demands explanation. Silence invites scrutiny.

This is not surveillance.

It is accountability evolving to match complexity.

The watcher notices that younger members of the organisation adapt to this posture more easily.

They have fewer habits built around episodic thinking. They are more comfortable holding ambiguity without resolution. For them, L-ARC™ does not feel like a framework–it feels like common sense finally given structure.

This generational ease is not accidental.

Institutions evolve not just through policy, but through who feels at home within them.

Another quiet consequence emerges: the institution's relationship with failure changes.

Failure is no longer defined solely by incidents or breaches. It is also defined by missed opportunities to act early when coherence was visible. This is a far more uncomfortable definition, because it implicates restraint as much as error.

But it is also a more honest one.

Honesty, in complex systems, is a form of resilience.

As Chapter 2 deepens toward its close, the institution becomes aware of a paradox it must now live with.

The better it becomes at anticipation, the harder it is to prove its value.

Success looks like nothing happening.

This paradox does not disappear.

Instead, it becomes part of governance itself.

Leaders learn to defend prevention without spectacle. Boards learn to evaluate foresight without certainty. Capital learns to wait without guarantees. These are not skills traditionally rewarded in institutional life.

But convergence has made them necessary.

The watcher reflects that the most radical outcome of this chapter is not technological at all.

It is the quiet erosion of the belief that certainty is a prerequisite for responsibility.

Responsibility, the institution has learned, begins when coherence appears–not when outcomes force agreement.

This lesson cannot be unlearned.

As the chapter draws toward its final resting place, the institution understands that it has crossed into a new category of system.

It is no longer merely reactive.

It is not yet predictive in the traditional sense.

It is anticipatory–and therefore accountable.

Accountability, once extended into the future, changes everything.

Chapter 2 ends not with confidence, but with resolve.

Resolve to design deliberately rather than drift.

Resolve to embed anticipation rather than rely on heroics.

Resolve to accept that early action will sometimes look unnecessary–and that explaining why it was taken will always matter more than celebrating its success.

The institution now knows that listening is not passive.

Listening is work.

The watcher, finally, is no longer watching alone.

The system watches with them.

And because it watches, it remembers.

And because it remembers, it must choose.

What Chapter 3 will confront is not whether this choice is desirable.

It is whether it can be operationalised without losing the moral clarity that made it necessary.

The work ahead is architectural, procedural, and political.

But the foundation is secure.

The institution has learned to hear the language of risk.

It has accepted the cost of understanding.

It has chosen responsibility over comfort.

What This Chapter Established

When AI Sees What Institutions Cannot Yet Act On

AI systems increasingly detect wildfire and ecological risk through weak signals, convergence, and environmental patterns. The intelligence is accurate. The warnings are early. Action still stalls. The constraint is no longer technology–it is legitimacy.

Executive Takeaways — What This Means for Leaders

- Risk does not begin with events. It begins with patterns that evolve across time and systems.
- AI does not create intelligence—it reveals persistence that institutions often normalise.
- Convergence is the most important signal in modern risk environments.
- The challenge is no longer detection. It is interpretation and escalation.
- Institutions must learn to act on coherence—not wait for confirmation.

What Comes Next

Why Trust Becomes the Limiting System

If AI can already perceive emerging fire and wildlife risk, the question then shifts from detection to authority. The next chapter explores why accurate intelligence fails without trust and why governance, not models, becomes the bottleneck.

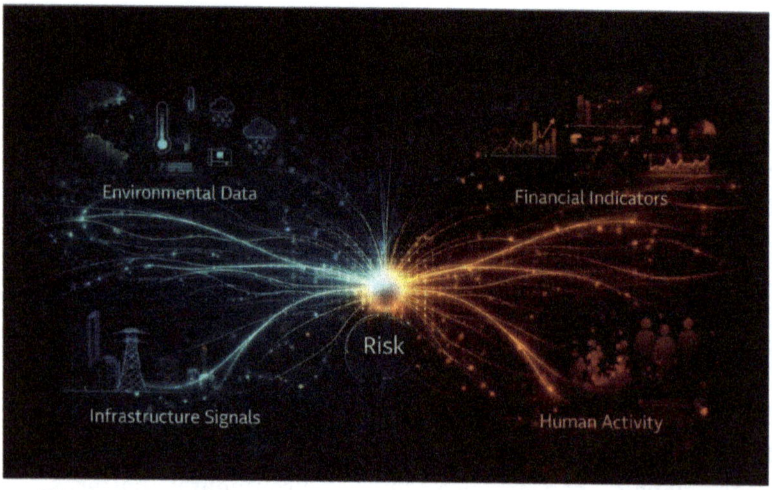

Figure 2.14 Risk is not created-it is the convergence of signals

Chapter 3

Trust Is the Hardest System to Build

Why Intelligence Fails Without Legitimacy

The institution does not doubt the intelligence anymore.

That is the quiet but decisive shift that separates this chapter from the last.

The signals are now visible.

The patterns have been named.

The convergence has been recognised.

What remains unresolved is something far more fragile.

Permission.

The watcher feels this before anyone says it aloud.

They see it in the way meetings now begin differently–with dashboards already open, not to search for problems, but to justify restraint. They hear it in the phrasing of questions that no longer ask whether the system is right, but whether acting on it can be defended. They notice how intelligence is acknowledged, even respected, yet still held at arm's length.

The system believes.

It simply does not trust itself enough to act.

The Model Everyone Trusted — Until It Mattered

A risk manager had relied on the model for months.

It had been right-quietly, consistently, without drawing attention.

It identified patterns earlier than anyone else could.
It explained its outputs clearly enough to be accepted.

Trust had formed.

Until the day it mattered.

The model flagged elevated risk-not extreme, not urgent, but rising.

The room went quiet.

No one questioned the intelligence.

But a different question emerged:
"Can we defend acting on this?"

No one challenged the model.

But no one moved either.

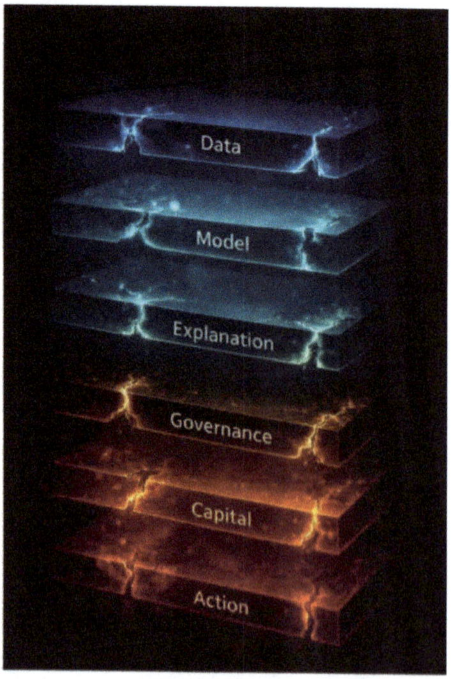

Figure3.1 The system does not fail at data-it fails at translation into action

This is the terrain of Chapter 3.

Because intelligence does not fail when it is wrong.

It fails when it is right–but illegitimate.

The Decision That Could Not Be Explained

A governance lead reviewed the recommendation.

The action itself was simple–adjust operations, reduce exposure, act early.

Nothing extreme. Nothing disruptive.

The problem was not the action.

It was the explanation.

"Why now?" someone asked.

The answer existed–in probabilities, correlations, and emerging patterns.

But not in a way that would withstand scrutiny later.

Not in a way that could be defended in a boardroom or an audit.

So the decision was delayed.

Not because it was wrong.

But because it could not yet be explained in the language the institution trusted.

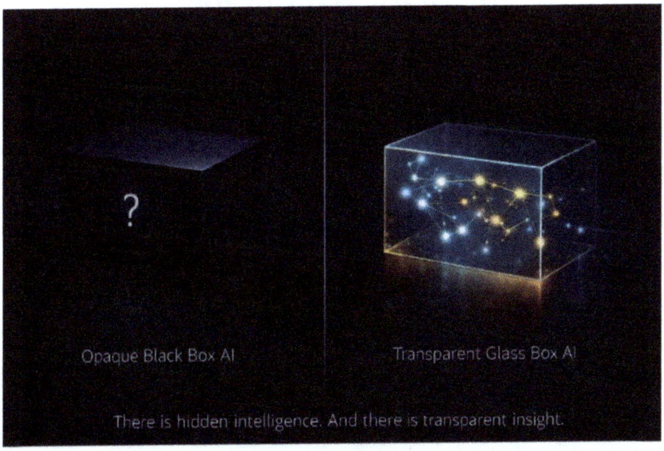

Figure 3.2 Unexplained intelligence is indistinguishable from risk

Figure 3.3 Trust as Infrastructure: Why Intelligence Waits and How Institutions Learn to Act Early

Accuracy Isn't Authority

Why Good Data Still Gets Ignored?

The institution is now saturated with insight.

Models run continuously.

Signals converge faster than reports can be written.

Near-interventions are no longer rare–they are expected. And yet, action remains cautious.

Not because the intelligence lacks accuracy, but because accuracy alone does not grant authority.

In wildfire and wildlife protection contexts, this gap carries unique weight. When institutions hesitate, ecosystems do not.

Animals relocate permanently, breeding cycles break, and recovery windows close. Trust delays here do not merely postpone action–they reshape landscapes in ways that governance cannot later undo.

The watcher watches a familiar pattern repeat itself, this time with better data.

A recommendation appears–clearer than before, supported by longer memory, reinforced by convergence across systems. It does not predict catastrophe. It does not overstate urgency. It does exactly what the institution once demanded: it is measured, explainable, defensible.

And still, it stalls.

Someone asks who owns the decision.

Someone else asks who signs if it goes wrong.

A third asks how this would look in hindsight.

None of these questions challenge the intelligence.

They challenge its legitimacy.

This is the first hard truth practitioners encounter when AI systems mature: being right is no longer enough.

Authority does not emerge from accuracy.

It emerges from alignment–legal, procedural, financial, and cultural.

The institution was designed to respond to events that justify action after the fact. It was not designed to act on intelligence that must be defended before anything happens.

So, the intelligence waits.

The invisible hierarchy of trust

What the watcher begins to understand is that trust inside institutions is layered.

Data may be trusted.

Models may be trusted.

Even predictions may be trusted.

But acting early requires a deeper form of trust–trust that spans governance, audit, capital, and personal accountability.

This trust has rules.

- Who is allowed to be wrong early
- What kinds of evidence count before harm
- Whether prevention can be justified if nothing happens
- Who absorbs reputational risk when success looks like nothing occurring

None of these rules are written in model documentation. They live elsewhere.

In policy manuals.

In board expectations.

In audit precedents.

In organisational memory of past overreactions.

This is why intelligence fails quietly. Not because it lacks credibility–but because credibility is not the same as permission.

A Vignette: When the Model is Believed But Ignored

A review session is convened.

The model has flagged a tightening corridor of risk–not imminent, but accelerating. The presentation is careful. Confidence intervals are shown. Limitations are stated explicitly.

No one disputes the analysis.

In fact, several participants nod.

One remarks that this aligns with what they've been sensing. Another notes that external indicators are beginning to echo the same concern. Then the conversation shifts.

What precedent would this set?

What if action is taken and nothing happens?

Would this survive scrutiny if challenged later?

The room is no longer evaluating intelligence.

It is evaluating legitimacy.

And legitimacy, unlike data, cannot be computed.

The recommendation is reframed as "input."

The action is postponed as "monitoring."

The intelligence is archived–not dismissed, but deferred.

The watcher recognises the pattern instantly.

This is not failure.

This is governance protecting itself.

The institution has reached a strange equilibrium.

Intelligence is no longer contested, but neither is it decisive.

This is the most dangerous state a system can enter–not ignorance, not denial, but polite acknowledgement without consequence. The watcher recognises it immediately. It feels orderly. Professional. Mature. It is also inert.

Accuracy has won the argument.

Authority has not been granted.

Inside the institution, there is an unspoken veto that operates without ever being exercised explicitly.

No one says no to the intelligence.

No one dismisses the data.

No one challenges the models.

Instead, action dissolves into process.

Requests for "additional validation."

Calls for "broader alignment."

References to "timing considerations."

Each request is reasonable in isolation. Collectively, they form a barrier that accuracy cannot cross.

The watcher understands now that this veto is not malicious. It is protective. It exists to shield individuals from personal exposure in systems that reward caution more reliably than foresight.

Accuracy threatens this equilibrium.

A Vignette: The Meeting Where Everyone Agrees

A steering group convenes. The presentation is crisp. The intelligence is strong. The convergence is clear. Several participants speak in support.

"This matches what we're seeing."

"This aligns with external indicators."

"We should keep a close eye on this."

Agreement fills the room.

And then–nothing happens.

The meeting concludes with a decision to "note" the analysis and revisit it at the next cycle. No one objects. No one advocates escalation. Consensus has been achieved without commitment.

Outside the room, the watcher reflects on the paradox: the more accurate the intelligence becomes, the easier it is to agree without acting. Disagreement forces decisions. Agreement allows deferral.

Institutions are not optimised for truth. They are optimised for defensibility.

This is not a flaw; it is a survival mechanism. Defensibility protects organisations from blame, litigation, political fallout, and reputational damage. It is deeply embedded in governance structures, performance metrics, and career incentives.

Accuracy, by contrast, is dangerous.

Accurate intelligence invites action before consequences justify it. It exposes decision- makers to retrospective scrutiny in worlds that still judge decisions by outcomes rather than reasoning.

The watcher has seen this before, long before AI entered the picture. Early warnings about structural risks, environmental pressures, or operational fragility were often acknowledged–but rarely acted upon–until something broke.

AI did not change this dynamic. It intensified it.

Acting early creates a unique form of vulnerability.

If the institution acts and nothing happens, then the action looks unnecessary. Wasteful. Overcautious.

If the institution waits and something happens, then the failure is attributed to circumstance, not decision.

This asymmetry explains why accuracy struggles to become authority. The watcher hears this logic articulated quietly, often indirectly:

"We can't be seen to overreact."

"We need something more concrete."

"Let's wait until we're sure."

Surety is the language of hindsight.

The institution knows this intellectually. But structurally, it remains bound to it.

Another obstacle emerges as intelligence improves: diffusion of responsibility.

When insights are produced by systems rather than individuals, ownership becomes ambiguous. Who stands behind the recommendation? Who answers, if questioned later?

The watcher notices how often intelligence is referred to as "the model's view," subtly distancing it from human judgment. This linguistic shift matters. It relocates accountability away from decision-makers and into abstraction.

Accuracy alone cannot bridge this gap.

Authority requires ownership.

A Vignette: When No One Signs

A proposal circulates–measured, reversible, defensible. It has passed technical review. Legal sees no immediate issue. Risk agrees it is prudent.

And yet, the proposal stalls.

Not because of opposition, but because no single role carries the mandate to authorise early action. Each function sees itself as advisory. Each expects another to decide.

The watcher realises that intelligence has outpaced institutional role design.

The system can see earlier than it can decide.

Paradoxically, improving accuracy often deepens the legitimacy gap.

As models become more nuanced, they surface complexity rather than certainty. They present probability distributions instead of thresholds, trajectories instead of triggers. This is intellectually honest–but institutionally destabilising.

Decision-makers accustomed to binary signals struggle to act on gradients.

The watcher sees frustration emerge–not with the models, but with what they demand. They demand judgment without guarantees. They demand accountability without closure.

This is where intelligence stalls most often.

Not at the edge of capability–but at the edge of courage.

The institution begins to confront a myth it has long upheld: that decisions are–or should be–data-driven.

In practice, decisions are legitimacy-driven.

Data informs.

Legitimacy authorises.

The watcher sees this distinction play out repeatedly. Data opens doors. Legitimacy decides whether anyone walks through them.

This realisation is unsettling, because it exposes a blind spot in many AI strategies. They assume that better data will naturally lead to better decisions. Chapter 3 reveals why this assumption fails.

Without legitimacy, intelligence accumulates but does not move.

In some quarters, highly accurate intelligence is experienced not as support, but as provocation.

It challenges existing narratives.

It disrupts comfortable timelines.

It exposes the cost of waiting.

The watcher observes subtle resistance–not denial, but deflection. Attention shifts to edge cases, uncertainties, and methodological debates. These are important discussions, but they often function as delay mechanisms.

Accuracy becomes something to be managed rather than acted upon.

By the end of this extended arc, the institution reaches a sobering insight: Accuracy is necessary, but authority is constructed elsewhere.

Authority is built through:

- governance pathways that pre-authorise early action
- audit frameworks that reward reasoning, not outcomes
- leadership norms that protect those who act before harm
- capital structures that tolerate prevention without spectacle

None of these emerge from models.

They must be designed.

The watcher stops thinking about how to make the intelligence louder.

They begin thinking about how to make the institution safer for those who act on it.

This is a profound shift.

It marks the transition from intelligence-building to trust-building.

And it is why the next section must move inside the machine–not to admire its sophistication, but to expose it to human systems of accountability.

The institution continues to behave as though the problem is technical.

That is its default posture.

When discomfort rises, it reaches for measures that feel objective: more tests, more validation, more process. It asks the model to prove itself again, because proofs are easier to request than permissions are to redesign.

The watcher sees the pattern clearly now.

Each time intelligence threatens to become actionable, the institution turns the request into a technical question, because technical questions have safe answers.

Permission questions do not.

A new cycle of validation begins.

Not because the previous validation failed, but because validation has become the institution's preferred form of delay. It is delay disguised as rigor, caution disguised as professionalism.

The watcher watches the same people who quietly believe the intelligence ask for "one more run," "one more comparison," "a stronger baseline."

The requests are not unreasonable.

But they are not the real reason action is not taken.

The real reason is that validation moves responsibility away from the decision and back into the machine. If the machine can be made to say something with absolute clarity, then acting becomes defensible without courage.

But the world rarely offers absolute clarity.

Persistence is rarely a headline.

Convergence is rarely a siren.

A Vignette: The Model Passes, and Still Nothing Moves

A formal review is held.

The model is evaluated against historical patterns. Its performance is strong. Its limitations are documented. Its confidence bounds are understood. The room is satisfied.

Then someone asks the question that always arrives eventually:

"So, who is prepared to act on this?"

The silence that follows is different from earlier silences.

Earlier, silence meant doubt. Now it means risk.

Not risk of disaster–risk of exposure.

The intelligence has become good enough that ignoring it no longer feels like neutral caution. It feels like a choice that someone might have to defend later.

But no one wants to be the first person to carry that defence alone.

The meeting ends with agreement to "socialise" the findings.

Socialising is the institution's way of distributing accountability so thinly that no single person owns it.

What the institution begins to realise–slowly and reluctantly–is that authority is not generated by accuracy.

Authority is generated socially.

It emerges from:

- repeated decisions that were defended successfully
- organisational memory that rewards early action rather than punishing it
- precedent that makes anticipation normal rather than exceptional
- structures that protect decision-makers when outcomes remain invisible

In other words, authority is an institutional asset that must be built.

And the institution does not yet have it.

The watcher sees this truth settle into discussions. It appears not as a declaration, but as a shift in tone. People begin to speak about "readiness." About "mandate." About "governance scaffolding."

These words are not technical. They are architectural.

Being right early creates enemies you never intended to make.

Not personal enemies, but institutional friction.

Early intelligence threatens budgets. It disrupts quarterly plans. It forces uncomfortable reallocations. It exposes fragile assumptions. It challenges leaders who have built credibility on the idea that their system is stable.

The watcher has watched this happen often enough to recognise it in small reactions:

A tightening of language.

A request to "reframe" urgency.

A suggestion to "tone down" what the model implies.

No one is lying.

They are protecting cohesion.

Institutions prioritise cohesion because cohesion keeps them functioning. But cohesion becomes dangerous when it requires denial of what is already known.

Accuracy collides with cohesion.

This is the politics of anticipation.

A Vignette: The Boardroom That Cannot Be Rushed

A board pack includes a new slide.

The intelligence is presented carefully: no dramatic claims, no fatalism. Just persistence, convergence, narrowing optionality. The language is executive-safe.

The watcher does not attend the board meeting, but they hear about it later. Not through gossip–through the subtle changes that follow.

The board is not sceptical. It is cautious in a different way.

It asks questions that reveal the real constraint:

"If we act on this, then what are we committing ourselves to?"

"What is the governance path for early action?"

"What does success look like if nothing happens?"

"Who will challenge us later, and on what grounds?"

These are not questions about accuracy.

They are questions about legitimacy.

The board does not fear being wrong.

It fears being undefended.

Even when no auditor is present, audit shapes behaviour.

This is one of the institution's most invisible forces.

The watcher begins to notice that people speak as though an audit is always listening. Decisions are framed not just for effectiveness, but for future review. Documentation becomes not just record-keeping, but defence-building.

AI complicates this.

Traditional audits are comfortable with events. They can trace decisions to incidents, thresholds, breaches. They are less comfortable with patterns, persistence, convergence–because these do not produce clean causal lines.

If the institution acts early, then auditors may later ask: "What triggered this?"

But convergence does not "trigger."

It accumulates.

So, the institution hesitates–not because it doubts reality, but because it doubts audit's ability to recognise responsibility when the world remains quiet.

The watcher realises something sobering:

The institution is not governed by risk alone.

It is governed by what it believes will be accepted as defensible later.

Audit is not just oversight. It is a design constraint.

A second near-intervention emerges, sharper than before.

It is proposed not as a dramatic intervention, but as a structured pre-commitment: a governance mechanism that authorises a defined set of early actions when convergence criteria are met.

This is important.

The institution is no longer trying to "be brave." It is trying to remove bravery from the system.

The proposal is careful:

- convergence thresholds defined (not numeric alone, but multi-domain coherence)
- explainability requirements specified
- accountability roles assigned
- financial envelope pre-approved
- audit trail templated

This is legitimacy engineered.

It is also threatening, because it changes power.

Pre-authorised early action reduces the control of those who benefit from delay. It reduces the power of committees that exist to slow decisions. It shifts authority from individuals to structure.

Structures redistribute risk.

Redistribution always provokes resistance.

The pushback arrives, but it is subtle.

It does not look like rejection.

It looks like caution about "governance complexity."

It looks like concern about "setting precedent."

It looks like a request to "pilot" the mechanism in a limited area.

Pilots are reasonable.

Pilots are also how institutions keep new authority small enough to remain controllable.

The watcher hears the phrase that always marks the moment power begins to feel threatened:

"Let's not move too fast."

This time, "too fast" does not mean operational risk.

It means governance reconfiguration.

In the background, the machine continues to do what it does best: remember.

It holds the convergence story steady. It does not allow the institution to revise its memory to fit comfort. It keeps producing the same coherence, the same persistence, the same narrowing optionality.

The watcher realises something that becomes a turning point of its own: The machine is not demanding action. It is refusing to provide forgetting.

For the institution, this is the most disruptive capability of all.

Because forgetting has always been the institution's safety valve.

When patterns persist without consequence, forgetting allows normalisation.

When forgetting is removed, normalisation becomes dishonest.

Dishonesty is unstable.

The climax of this extended section is not an event.

It is a sentence spoken in a room where decisions are supposed to be made.

Someone says, quietly, without accusation: "The model isn't the problem. We are."

The room goes still.

Not defensive stillness.

Recognition stillness.

Everyone understands what the sentence means:

- Accuracy is sufficient
- Convergence is real
- Memory is holding
- Capital is already shifting
- The missing component is legitimacy architecture

The institution has reached the edge of its own design.

This is the true ending of this Section.

It is the point where the story can no longer pretend that better intelligence will solve the problem. The problem is trust–and trust is not a model property.

Trust is a system. And that system must now be built.

Case Study

When the Machine Was Right — Until It Wasn't: Google Flu Trends

Google once tried to predict flu outbreaks using search data.

At first, it worked. The system picked up patterns faster than traditional reporting. It felt like a breakthrough–machines noticing what humans could not.

Then it started getting it wrong.

Search behaviour changed. Media influenced people. The model kept producing confident outputs, but drifted further from reality.

Trust faded quickly.

The issue wasn't just accuracy.

It was understanding when not to trust the machine.

AI can detect patterns early.

But without trust and context, those patterns don't translate into decisions.

Executive Takeaway:

AI without trust is intelligence without authority.

Seeing Inside the Machine

Explainability, Accountability and Human Judgement

The institution does not fear intelligence. It fears opacity paired with responsibility.

This fear is older than machines. Long before algorithms entered decision-making, institutions struggled with decisions whose logic could not be fully narrated after the fact. What machines have done is make that discomfort unavoidable.

The watcher notices the shift the moment intelligence becomes persistent enough to matter.

Earlier, when signals were sporadic, opacity could be tolerated. Ambiguity could be absorbed into judgment. Decisions could be explained by reference to experience, instinct, or precedent. But as machine memory stabilises patterns over time, decisions informed by it acquire a different character.

They are no longer moments. They are positions.

And positions must be defensible.

One of the institution's earliest misunderstandings is subtle but consequential. It assumes that opacity means the machine cannot explain itself.

In practice, the machine can produce explanations endlessly: feature weights, attribution maps, confidence bands, causal traces. None of these satisfy the institution's deeper need.

Because what the institution requires is not explanation. It requires narrative accountability.

The watcher realises this during a review session that feels strangely unproductive.

The model team presents its work. Visualisations are clear. Assumptions are documented. Limitations are acknowledged. By any technical standard, the system is explainable.

Yet the room remains uneasy.

Someone finally articulates the discomfort:

"I understand how it works. I don't yet understand how I would defend acting on it."

That sentence becomes a quiet refrain across the institution.

Explainability, the watcher learns, is never a single requirement.

It is plural.

What satisfies a data scientist does not satisfy a regulator.

What satisfies a regulator does not satisfy a board.

What satisfies a board does not satisfy an inquiry held years later, under different political and social conditions.

The institution begins to recognise that explainability must travel.

It must survive translation across time, roles, and scrutiny.

This is where many AI initiatives fail–not because they lack transparency, but because they misunderstand the audience of explanation.

The watcher sees explainability discussions fragment:

- Technical explainability for developers
- Legal explainability for compliance
- Strategic explainability for executives
- Moral explainability for the public

Each demands a different story.

None can be ignored.

A Vignette: The Audit Room

An audit review is convened.

The auditors are not hostile. They are professional, thorough, and curious. They ask how decisions informed by the system are documented. They ask how thresholds are defined. They ask how exceptions are handled.

Then they ask the question that tightens the room:

"If this system influenced an early action, then how would you demonstrate that the action was reasonable at the time it was taken?"

The watcher notices the shift.

This is no longer about whether the system works.

It is about whether the institution can tell a story that survives hindsight.

The machine can explain what it saw.

The institution must explain why it trusted it enough to act.

That explanation does not yet exist.

There is a persistent myth that AI removes human judgment.

What Chapter 3 exposes is the opposite.

AI makes human judgment unavoidable.

When decisions were reactive, judgment could hide behind events. When harm occurred, response was justified by necessity. But early action strips away that shield.

Judgment becomes explicit.

The watcher sees this discomfort surface repeatedly.

Leaders are asked not just to decide, but to own the timing of their decisions. They must justify why they acted before harm, not after. They must articulate the values that guided restraint or intervention.

Machines cannot do this.

They can inform judgment. They cannot replace it.

As intelligence matures, a gap opens between insight and accountability.

The watcher watches this gap widen.

On one side, machine memory accumulates evidence steadily. On the other, institutional accountability remains episodic, triggered by events.

This mismatch creates paralysis.

No one disputes the intelligence.

No one claims ignorance.

Yet no one feels authorised to act.

The institution has clarity without courage. And courage, it turns out, is not a personal trait.

It is a structural one.

A Vignette: Legal Review Without Refusal

A legal team reviews a proposed early action informed by machine convergence.

They do not reject it.

They do not warn against it.

They do something more destabilising.

They ask for clarification.

Under what authority would this action be taken?

What precedent does it establish?

How would proportionality be assessed, if challenged later?

Each question is reasonable.

Each one exposes a missing structure.

The watcher realises that legality is not blocking action.

Undefined legitimacy is.

The institution initially believes that better explainability will unlock action.

More transparency.

More interpretability.

More documentation.

These efforts help.

They do not solve the problem.

Because legitimacy is not produced by understanding alone. It is produced by authorisation.

Explainability answers the question: How does this work?

Legitimacy answers a different one: Who is allowed to act on this, and under what conditions?

The institution has answered the first.

It has not yet designed the second.

One of the most uncomfortable consequences of anticipatory intelligence is that it personalises responsibility.

The watcher sees this most clearly in senior leaders.

When action follows an event, responsibility is shared with circumstance. When action precedes harm, responsibility sits squarely with the decision-maker.

This changes behaviour.

Leaders become cautious not because they doubt the intelligence, but because they understand the exposure it creates. Acting early requires not just confidence in the system, but confidence that the institution will stand behind the decision if challenged.

Trust shifts from technology to organisation.

A Second Vignette: The Inquiry that Hasn't Happened Yet

In a planning session, someone asks an uncomfortable hypothetical:

"If this were reviewed five years from now, then how would we explain why we acted?"

The room grows quiet.

This imagined future inquiry shapes the present more than any immediate risk. Decisions are filtered through a lens of retrospective judgment that has not yet occurred but is already powerful.

The watcher understands that institutions are governed as much by imagined futures as by present realities.

Explainability must travel forward in time.

As the institution struggles with legitimacy, the role of the machine subtly changes.

It stops being framed as a decision-maker.

It becomes a witness.

A witness that remembers what humans forget.

A witness that records convergence without emotion.

A witness that cannot be persuaded by comfort or convenience.

This reframing matters.

It allows the institution to reclaim judgment without losing insight.

The watcher sees language shift.

People stop saying "the model decided."

They begin saying "the system observed."

This linguistic change is not cosmetic.

It repositions responsibility.

The institution reaches a difficult conclusion:

Trust will not emerge organically.

It must be designed.

This means designing:

- governance pathways that define who can act early
- documentation standards that defend reasoning, not outcomes
- audit frameworks that recognise persistence as evidence
- leadership norms that reward prevention without spectacle

None of these are technical problems. They are institutional design problems.

A Vignette: Designing Legitimacy on Paper

A working group is formed–not to improve the model, but to redesign decision pathways.

This is new.

The conversation is uncomfortable.

Who should be authorised to act on convergence?

What thresholds are meaningful when patterns replace events?

How should dissent be recorded?

How should early success be measured?

The watcher listens carefully.

For the first time, the institution is not asking the machine to justify itself.

It is asking how to justify believing the machine early enough to matter. This is progress.

Confidence, the watcher realises, is not belief.

It is preparedness.

Confidence comes from knowing that:

- decisions will be defended even if outcomes are quiet
- accountability is shared, not isolated
- governance has anticipated scrutiny
- capital has accepted waiting

When these conditions exist, intelligence can finally move.

Until then, it waits.

The climax does not arrive with a breakthrough.

It arrives with a sentence spoken by someone who has never spoken this way before:

"We need to stop asking whether the machine can explain itself–and start asking whether we have explained ourselves well enough to act."

The sentence lands heavily.

It reframes everything.

Explainability is no longer a feature of the system.

It is a property of governance.

The machine has done its job.

Now the institution must do its own.

Once the institution accepts that explainability is not a technical property but a social one, the problem does not become easier.

It becomes heavier.

Because what must now be explained is not the machine.

It is the institution's own reasoning.

Institutions are accustomed to explaining actions taken under pressure.

They are far less comfortable explaining actions taken under foresight.

Pressure legitimises urgency.

Foresight exposes intention.

The watcher sees how this difference unsettles even experienced leaders. When decisions are reactive, responsibility is shared with circumstance. When decisions are anticipatory, responsibility feels deliberate, even elective.

The institution begins to realise that early action removes the shield of inevitability.

Nothing made the institution act. It chose to.

Choice demands explanation.

A Vignette: The Moment Explanation Turns Inward

In a senior review, a leader asks for a briefing on explainability.

The briefing begins with diagrams, feature attributions, and confidence bands. The leader listens politely, then interrupts.

"I don't need to know how the system works," they say.

"I need to know how we would explain acting on it."

The room falls quiet.

This is not a technical request. It is a governance one.

The watcher notices how the conversation shifts immediately. Slides are skipped. Language changes. People begin to speak about precedent, mandate, narrative, and defence.

Explainability has crossed its real threshold.

It has become introspective.

Early action creates a narrative vacuum.

When nothing happens, the institution must still tell a story.

This is profoundly uncomfortable for systems trained to justify action through outcomes. When outcomes are absent, reasoning must carry the full weight of legitimacy.

The watcher recognises this as the hidden cost of prevention.

Prevention denies institutions the evidence they traditionally rely on to prove they were right.

No incident. No damage.

No clear counterfactual.

Just a choice made early and the quiet that followed.

The institution does not yet know how to narrate quiet.

As anticipatory decisions become possible, audit transforms from a process into a presence.

Even when no audit is scheduled, its imagined gaze shapes behaviour.

The watcher notices this in the way documents are written. Language becomes careful, conservative, pre-emptively defensive. Rationales are layered not for clarity, but for survivability.

Decisions are explained as though someone hostile will read them later.

This imagined hostility is powerful.

It disciplines foresight more effectively than any formal rule.

Explainability, in this context, is less about transparency and more about future-proofing.

A Vignette: Explaining the Absence of Harm

A draft decision rationale is circulated.

It explains why a set of early actions were taken. The language is careful, restrained. It references convergence, persistence, and narrowing options.

Someone raises a concern:

"But how do we prove this mattered?"

The question is not cynical. It is existential.

The institution has no vocabulary for explaining success that looks like absence. There is no template for defending actions whose value lies in what did not occur.

The watcher realises that explainability is not missing. Language is.

At this stage, the institution begins to ask the machine for help with explanation.

Can it generate narratives?

Can it summarise reasoning?

Can it quantify avoided risk?

These capabilities help. They do not solve the problem.

Because the hardest part of explanation is not what was seen.

It is why that seeing was enough to act.

Machines cannot grant permission. They can only inform judgment.

As explainability becomes central, human judgment is no longer invisible. It is scrutinised.

The watcher notices how this scrutiny changes behaviour. Decision-makers begin to second-guess not the intelligence, but themselves. They wonder how their judgment will be perceived later.

This creates a new form of risk: reputational anticipation.

Leaders imagine how their decisions will be interpreted under different political, social, or regulatory climates. They worry less about being wrong and more about being misunderstood.

Explainability becomes a defensive art.

A Vignette: The Decision that Almost Collapses Under Explanation

A near-intervention reaches final review.

Everything is in place. Governance pathways are defined.

Accountability is shared. Capital is allocated.

Then the question arises:

"Can we explain this simply?"

The simplicity test is devastating.

The reasoning is sound, but complex. The explanation is accurate, but layered. The institution fears that complexity will be mistaken for uncertainty.

The decision wavers.

Not because it is wrong–but because it is hard to narrate cleanly.

The watcher sees this as a critical inflection point.

Institutions that cannot explain complexity will default to simplicity–even when simplicity is misleading.

Explainability redistributes power.

Those who control the narrative control legitimacy.

As the institution moves toward anticipatory action, new tensions emerge. Legal teams, risk functions, communications units, and executives all recognise that explanation is now a strategic asset.

Who gets to frame the story?

Whose language becomes authoritative?

Which explanations are acceptable?

These questions surface quietly, but they matter.

The watcher realises that explainability is becoming a contested space. Not because people disagree on the intelligence. But because they disagree on how much exposure they are willing to tolerate.

In this contested environment, the machine acquires a new role.

It becomes a stabiliser. Not of decisions–but of memory.

It holds the convergence steady while human narratives fluctuate. It prevents selective recall. It resists reinterpretation.

This is uncomfortable.

The institution is accustomed to revising its understanding of the past to fit present comfort. Machine memory limits that flexibility.

Explainability, paradoxically, increases discomfort by reducing narrative manoeuvrability.

Truth becomes less negotiable.

A Vignette: When Memory Refuses to Bend

In a later discussion, someone suggests that earlier signals were less clear than they seemed.

The machine's record is reviewed.

The signals were persistent.

The convergence was real.

The hesitation was documented.

The room grows quiet.

No accusation is made.

But the past no longer shifts easily.

The watcher recognises that explainability has crossed another threshold.

It is no longer about making decisions understandable. It is about making forgetting difficult.

At this depth, explainability becomes ethical.

It constrains the institution from convenient amnesia. It forces continuity of reasoning across time. It links early awareness to later accountability.

This is not punitive. It is stabilising.

The institution begins to understand that trust does not come from perfect decisions, but from coherent ones–decisions that can be traced, explained, and defended even when circumstances change.

Explainability is no longer an obligation imposed from outside.

It becomes a discipline the institution imposes on itself.

As this discipline takes hold, behaviour begins to change subtly.

Decision papers read differently.

Meetings feel different.

Silences mean different things.

People begin to ask not only whether a decision is justified, but whether it will remain justifiable.

The watcher sees this as the birth of institutional maturity.

Trust is no longer something to be claimed.

It is something to be sustained.

The climax of this extended section is diffuse rather than dramatic. It unfolds as a shared recognition across roles:

- ➢ The machine does not need to be trusted blindly
- ➢ The machine does not need to decide
- ➢ The institution needs to decide how it will explain itself when it acts early

This recognition dissolves many earlier tensions.

The problem was never opacity.

It was exposure without preparation.

Explainability, properly understood, is preparation.

By the end of this long interior movement, the institution understands something irreversible:

If it cannot explain its reasoning before crisis, then it does not deserve to act before crisis.

This is not a technical requirement. It is a moral one.

With explainability reframed as institutional self-knowledge, the final question of Chapter 3 becomes unavoidable:

How do you design governance that earns confidence before it is tested?

Designing Confidence Before Crisis

Governance that Earns the Right to Act

By the time the institution arrives here, it no longer believes that trust will emerge naturally.

That belief belonged to an earlier era–one in which authority was assumed, events were discrete, and responsibility could be located after the fact. Chapter 3 has already stripped that comfort away.

Trust, the institution now understands, must be designed.

Not asserted.

Not improvised.

Not defended after harm.

Designed before it is needed.

The watcher senses this shift immediately. It appears not in grand declarations, but in a quiet change of posture. Meetings are no longer framed around what the institution can do, but around what it has the right to do early–and under what conditions.

This is not hesitation. It is preparation.

One of the institution's most persistent misunderstandings is that early action requires courage.

Courage implies personal risk.

Heroism.

Exceptional behaviour.

But institutions cannot rely on courage. They require repeatability.

What the watcher begins to see is that the most anticipatory organisations are not braver. They are calmer. They do not ask individuals to take leaps of faith. They design systems in which early action feels ordinary–because it is authorised, documented, and expected.

Confidence, in this sense, is structural.

A Vignette: When Early Action No Longer Feels Exceptional

A governance meeting convenes to review a proposed adjustment informed by convergence. The adjustment is modest. It sits well within budget. It has been pre-authorised under defined conditions.

No one frames it as extraordinary.

No one asks who will be blamed if nothing happens.

The decision passes quietly.

The watcher realises something profound has changed.

Early action has become routine.

This is what confidence looks like inside institutions.

Designing confidence before crisis requires a shift in how legitimacy is granted.

Traditionally, legitimacy arrives through consequence. Events justify action. Harm confers permission. Authority is retroactive.

Anticipatory systems invert this logic.

They require pre-legitimacy–structures that define in advance:

- what constitutes sufficient early evidence
- who is authorised to act on it
- how dissent is recorded and respected
- how success is evaluated when outcomes are absent

The institution begins to build these structures deliberately.

Not as controls, but as scaffolding.

The watcher has long observed governance as a braking mechanism. Policies slowed decisions. Committees absorbed momentum. Review cycles deferred action.

In anticipatory systems, governance must perform a different role.

It must grant permission.

This requires rethinking governance not as protection from error, but as protection for those who act responsibly under uncertainty.

The institution begins to articulate this explicitly.

Governance is no longer framed as preventing mistakes.

It is framed as defending reasoning.

A Vignette: Redefining Escalation

An escalation pathway is redesigned.

Previously, escalation required thresholds to be breached. Now, escalation is permitted when defined patterns converge across domains–even if no single metric crosses a line.

This is uncomfortable.

Thresholds felt objective.

Patterns feel interpretive.

But the institution recognises that objectivity delayed action while interpretation can be defended–if designed carefully.

Escalation becomes a matter of coherence, not crisis.

One of the most counterintuitive insights of this section is that trust is strengthened–not weakened–by structured dissent.

The watcher notices how early governance designs deliberately preserve disagreement. Alternative interpretations are recorded. Minority views are documented.

This is not indecision. It is insurance.

When early action is later scrutinised, dissent demonstrates that judgment was exercised, not bypassed. It shows that the institution did not blindly follow intelligence, but weighed it responsibly.

Trust thrives in transparency of disagreement.

Confidence before crisis cannot be designed without capital alignment.

This is where many anticipatory efforts fail.

Finance is accustomed to funding responses to events. Prevention challenges traditional return logic. When nothing happens, value is hard to demonstrate.

The institution begins to redesign how capital supports anticipation.

Budgets are allocated not for outcomes, but for option preservation. Financial models begin to value flexibility, resilience, and reduced volatility.

The watcher hears finance leaders use new language:

"We're paying for stability, not savings."

"We're buying time."

"We're reducing future exposure."

This language matters.

It legitimises waiting as a strategic choice rather than a sunk cost.

A Vignette: Explaining Prevention to Capital

In a capital review, a question arises:

"How do we measure success if nothing happens?"

The response is calm.

"We measure whether options remained open."

"Whether escalation paths stayed reversible."

"Whether volatility was reduced."

No one mentions avoided disaster.

The institution has learned that prevention must be explained in terms capital understands.

Stability is a return.

Designing confidence before crisis cannot remain internal.

As convergence increases, institutions become interdependent. Early action by one affects other. Legitimacy must therefore be shareable.

The watcher observes cross-agency discussions shift tone. Instead of debating jurisdiction after events, agencies negotiate legitimacy before action.

They align on:

- shared interpretation frameworks
- mutual recognition of evidence
- coordinated escalation thresholds

This alignment does not remove friction.

It makes friction navigable.

Trust becomes collective.

Eventually, the institution confronts a final audience it cannot control.

The public.

Early action is difficult to explain externally. When nothing happens, scepticism emerges. Accusations of overreach, waste, or hidden agendas are easy to make.

The institution realises that public trust cannot be managed reactively. It must be cultivated.

Narratives are prepared–not defensive ones, but educational ones. The institution explains how it thinks, not just what it does.

It communicates uncertainty honestly. It frames early action as responsibility, not alarmism.

The watcher sees communications become part of governance, not an afterthought.

A Vignette: Speaking Before Harm

A spokesperson is briefed–not for crisis, but for calm explanation.

They explain why certain measures were taken. They speak about patterns, persistence, and preparation. They do not promise certainty.

The reaction is mixed.

But something important happens.

The institution speaks before it is forced to.

This is confidence.

The climax of this section arrives not as an action, but as a declaration embedded quietly in policy language:

"Early action taken under defined conditions will be presumed legitimate unless evidence of negligence is shown."

This sentence changes everything.

It reverses the burden of proof.

No longer must early action justify itself endlessly.

Now inaction must be explained.

The watcher recognises this as the moment trust becomes structural.

By the end of Chapter 3, the institution has changed in a way that is difficult to summarise but impossible to miss.

It no longer waits for certainty.

It no longer fears quiet outcomes.

It no longer treats trust as a feeling.

Trust has become infrastructure.

Designed.

Maintained.

Defended.

Designing confidence is not a single intervention. It is a sustained act of institutional self-discipline.

Once the initial structures are put in place–pre-legitimised escalation pathways, shared accountability, capital alignment–the institution discovers that design alone is not enough. Confidence must be maintained under pressure, ambiguity, and fatigue.

This is where most systems fail.

Not at conception.

At endurance.

Confidence, unlike fear, erodes quietly.

The watcher notices this in the months that follow early success. No crisis occurs. The system remains stable. The early actions taken appear to have mattered–but there is no dramatic proof.

This quietness becomes dangerous.

People begin to ask, gently, whether the structures are still necessary. Whether the overhead is justified. Whether the institution has overcorrected.

The very success of anticipation invites complacency.

The institution must now govern its own relief.

A Vignette: When Prevention Starts to Feel Wasteful

A budget review revisits anticipatory investments.

Nothing bad has happened.

This is framed as good news–but it subtly reframes prevention as excess. Questions are asked about opportunity cost. About efficiency. About whether resources could be reallocated now that conditions appear stable.

No one proposes dismantling governance outright.

They propose "optimisation."

The watcher recognises optimisation as the first stage of forgetting.

To survive its own success, confidence must become ritual.

Not symbolic ritual–but operational repetition.

The institution begins to formalise practices that keep anticipation alive even when risk appears distant:

- periodic revalidation of convergence criteria
- scheduled near-intervention reviews, even when none occur
- governance simulations that rehearse early action
- audit walkthroughs of decisions that were not taken

These rituals do not respond to danger. They maintain readiness.

The watcher realises that anticipation is not sustained by vigilance alone, but by habit.

One of the most fragile elements of designed confidence is memory.

People move roles. Leaders rotate. Priorities shift. Without deliberate preservation, the reasons early governance was built begin to fade.

The institution begins to treat memory as infrastructure.

Decisions are archived not just as records, but as narratives. Why early action was taken. Why restraint was chosen. What alternatives were considered.

This narrative memory becomes part of onboarding, leadership transition, and governance review.

The machine holds data.

The institution must hold meaning.

A Vignette: Leadership Change Without Collapse

A senior leader leaves.

Their successor inherits anticipatory governance structures they did not design. This is a critical test.

The watcher watches closely.

Instead of dismantling what they did not create, the new leader asks to be walked through the reasoning history. Not just the decisions, but the logic that justified early action before crisis.

The handover is slower than usual. But nothing collapses.

This is a rare success.

Confidence has outlived its architects.

Designed confidence eventually attracts political attention.

Early action is visible. Budgets shift. Measures are taken without public incidents to justify them.

The institution begins to experience external pressure:

- media scrutiny
- stakeholder questioning
- political discomfort with action absent crisis

This pressure is inevitable.

The watcher observes how the institution responds.

Not defensively. Deliberately.

It publishes reasoning frameworks. It explains how legitimacy is earned. It invites scrutiny of process rather than secrecy of outcomes.

This openness is risky. It is also stabilising.

Confidence cannot survive secrecy.

A Vignette: Explaining Restraint Publicly

At a public briefing, a journalist asks: "What exactly were you preventing?"

The spokesperson does not speculate.

They explain how the institution thinks about patterns, convergence, and preparation. They explain that some actions are taken to preserve options, not to avert named disasters.

The answer is unsatisfying to those who want certainty.

It is honest.

The watcher understands that honesty is the only sustainable posture.

Over time, fatigue emerges. Not from crisis–but from vigilance.

People tire of holding ambiguity. Of explaining decisions that look unnecessary. Of defending budgets without dramatic proof.

Fatigue is the most underestimated risk in anticipatory systems.

The institution begins to design for it.

Rotations are planned. Decision burdens are shared. No one role is allowed to carry foresight alone.

Trust is distributed.

A Vignette: Rotating Responsibility

A governance role responsible for early escalation is rotated deliberately.

The purpose is not redundancy. It is empathy.

Each participant learns what it feels like to own foresight. To defend quiet decisions. To explain absence.

This shared experience strengthens institutional trust more than any policy.

Capital, too, must be protected from fatigue.

Investors and funders tolerate waiting only if they believe the institution knows why it waits.

The institution learns to report anticipation differently:

- not as avoided loss
- but as preserved stability
- not as risk reduction
- but as volatility control

Over time, capital adapts.

Patience becomes priced.

This is rare.

It is also decisive.

The second climax arrives not with resistance–but with temptation.

Conditions improve.

Signals soften.

Someone suggests scaling back anticipatory posture.

The institution pauses. This pause matters.

It reviews its own criteria–calmly, deliberately. It does not cling to vigilance out of fear. It does not dismantle structures out of relief.

It adjusts without abandoning.

The watcher recognises this as maturity.

By the true end of Section 3.3, the institution has learned its hardest lesson: Confidence before crisis is not about permanent alertness. It is about adaptive legitimacy.

The ability to tighten and loosen posture without losing coherence.

This is what distinguishes resilient institutions from anxious ones.

Intelligence made anticipation possible.

Explainability made trust discussable.

Governance made confidence repeatable.

But only design made it durable.

Chapter 3 closes with an institution that no longer asks whether it should trust itself. It has learned how.

The institution can now see.

It can now trust.

It can now act early.

What remains is sustainability.

How does this posture survive years, not moments?

How does it scale beyond single organisations?

How does it become economic, not exceptional?

That is the next challenge.

By the time the institution reaches the end of this chapter, it no longer speaks of trust as an aspiration.

Trust has become infrastructure.

It is designed, maintained, stress-tested, and governed. It no longer lives in sentiment, reputation, or individual confidence. It lives in process, permission, and preparedness.

This is the most important outcome of Chapter 3.

Because intelligence without trust does not fail loudly.

It fails quietly.

For most of its history, the institution believed that trust preceded action. That leaders had to feel confident before they could act.

Chapter 3 reverses that logic.

Trust is not the cause of early action. Trust is the product of well-designed action taken under ambiguity.

This inversion matters because it shifts responsibility away from psychology and into architecture. It removes the expectation that individuals must be brave enough to act early and replaces it with a requirement that systems must be safe enough for early action to be ordinary.

The institution no longer asks whether people trust the intelligence. It asks whether the institution has earned the right to trust itself.

Chapter 3 makes explicit something that was previously implicit.

Intelligence was never waiting because it lacked accuracy. It was waiting because the institution lacked legitimacy pathways that could carry foresight without personal exposure.

Accuracy produced insight.

Explainability produced understanding.

But neither produced permission.

Permission was missing. And permission, it turns out, is the rarest institutional resource of all.

What practitioners often underestimate is how many layers trust must traverse before it becomes actionable.

Trust must exist:

- at the operational layer, where signals are observed
- at the governance layer, where decisions are authorised
- at the audit layer, where actions are later examined
- at the capital layer, where patience must be funded
- at the public layer, where legitimacy must be sustained

If trust collapses at any one of these layers, intelligence stalls.

Chapter 3 demonstrates that trust is not a single relationship. It is a system of aligned permissions.

One of the quiet warnings embedded in this chapter is that confidence without design is dangerous.

Institutions sometimes mistake decisiveness for trust. They act quickly, boldly, and visibly–believing that speed itself confers legitimacy.

Chapter 3 shows why this is false.

Unstructured early action creates fragility. It concentrates risk, invites backlash, and erodes confidence over time. When outcomes remain quiet, unearned confidence collapses into embarrassment or retreat.

Trust must be earned slowly to be exercised quickly.

Chapter 3 also resolves a common misunderstanding about explainability.

Explainability is not about opening the black box. It is about opening the institution.

The decisive question was never: *Can the machine explain itself?*

It was always: *Can the institution explain why it chose to act early–and stand by that explanation later?*

This reframing shifts investment away from purely technical interpretability and toward institutional self-knowledge. Documentation, narrative discipline, and reasoning continuity become as important as model transparency.

Explainability becomes a mirror, not a microscope.

Contrary to popular fears, Chapter 3 does not depict a future where machines replace judgment.

It depicts a future where judgment becomes unavoidable.

When action precedes harm, human values are exposed. Priorities, tolerances, and trade-offs must be articulated explicitly. Judgment can no longer hide behind events.

This exposure is uncomfortable.

It is also essential.

Institutions that cannot tolerate exposed judgment will always wait for crisis.

Perhaps the most practitioner-relevant insight of Chapter 3 is the redefinition of governance.

Governance has long been framed as restraint–something that slows action to prevent mistakes.

In anticipatory systems, governance must perform a different function. It must enable early action by pre-legitimising it.

This requires:

- escalation criteria based on convergence, not thresholds
- documented dissent rather than forced consensus
- shared accountability rather than isolated decision-making
- presumptive legitimacy for early action unless negligence is shown

This is governance that earns trust rather than policing it.

Trust cannot be sustained without capital alignment.

Chapter 3 demonstrates that anticipatory governance fails when finance remains reactive. If capital only rewards visible outcomes, then institutions will drift back toward event-driven behaviour.

The shift described here is subtle but profound.

Capital begins to value:

- volatility reduction over cost savings
- option preservation over immediate return
- endurance over optimisation

When capital learns to wait, trust becomes economically viable.

Without this alignment, anticipation remains performative.

One of the most underappreciated insights of this chapter is psychological.

Waiting is not neutral.

It feels safe.

It feels prudent.

It feels ethical.

But when intelligence is clear and convergence persistent, waiting becomes a decision with consequences.

Chapter 3 reframes waiting not as restraint, but as an active choice that must be defended like any other.

This reframing changes conversations fundamentally.

Inaction must now explain itself.

By the end of Chapter 3, the watcher has changed. Not in status or authority–but in necessity.

Their unease is no longer singular. It has been absorbed into process. Their intuition has been translated into criteria. Their vigilance has been distributed across systems.

The watcher does not disappear.

They are no longer alone.

This is the quiet success of institutional trust.

Perhaps the hardest truth in this chapter is that trust cannot be proven in moments.

It is proven over years.

Through leadership transitions.

Through political cycles.

Through budget pressures.

Through periods when nothing happens and questions grow louder.

Chapter 3 shows that designed trust survives these tests–not because it is rigid, but because it is adaptive.

Structures tighten and loosen without collapse.

Confidence breathes.

The institution that emerges from Chapter 3 is not heroic.

It is mature.

It does not rely on individuals to take risks on its behalf. It does not wait for certainty to absolve it of responsibility. It does not confuse caution with ethics.

It has learned to design trust deliberately.

That design is its quiet advantage.

With trust now structured, explainable, and economically supported, the institution faces its final challenge: How to sustain this posture at scale–across systems, sectors, and time.

That is not a question of intelligence. It is a question of continuity.

And it is where the book must now go.

What This Chapter Established

Trust as Infrastructure, Not Instinct

Trust failed not because leaders were unwilling, but because institutions lacked structures that grant early intelligence the right to act. Accuracy without legitimacy leaves foresight advisory and optional. Trust, when it matters most, must be engineered.

Executive Takeaways — What This Means for Leaders

- Accuracy does not create action. Trust does.
- Explainability is not a technical feature–it is a governance requirement.
- Intelligence without legitimacy remains unused.
- Institutions act when decisions are defensible–not when they are merely correct.
- Building trust in early signals is the most critical system design challenge.

What Comes Next

Turning Legitimacy into Capital

If trust creates permission, then permission must eventually move money. The next chapter examines how prevention becomes fundable and how institutions learn to value what does not produce visible loss.

Chapter 4

The Economics of Not Burning

Making Prevention Make Financial Sense

After Trust Comes the Invoice

The institution has learned to listen.

That is the quiet victory of the previous chapter.

Signals are no longer dismissed.

Patterns are no longer invisible.

Governance no longer waits blindly for catastrophe.

Trust has been designed.

And now, a different question surfaces–one that feels less philosophical and far more uncomfortable.

Who Pays for Acting Early?

The watcher notices the shift immediately.

The conversations have changed tone. They no longer revolve around whether the intelligence is credible, or whether early action is legitimate. Those debates have been settled–carefully, structurally, deliberately.

Now the questions arrive from a different direction.

They arrive with spreadsheets.

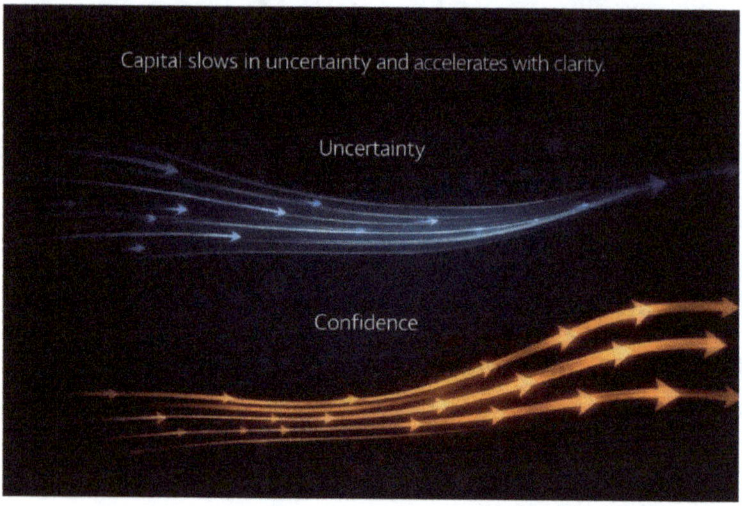

Figure 4.1 Capital does not fear risk–it fears uncertainty

They arrive with budget cycles.

They arrive with phrases like "business case," "opportunity cost," and "return profile."

This is where many prevention efforts quietly fail.

Not because leaders disagree with the logic–but because the economics have not been rewritten to match the new reality.

The Paradox of Prevention

In the weeks following a near-intervention that does not escalate into crisis, the institution feels something close to relief.

Nothing happened.

No damage.

No headlines.

No emergency expenditure.

Figure 4.2 The greatest success is the disaster that never happens

From the outside, the decision to act early looks indistinguishable from overreaction.

Inside the finance function, this absence becomes a problem.

There is no loss to quantify.

The watcher hears this logic second-hand, translated through layers of governance:

"We did the right thing–but how do we justify funding this again?"

This is the central economic failure of prevention.

The Budget That Could Not See Success

In the finance review meeting, the system was discussed calmly.

Early signals had been detected.
Preventive actions had been taken.

Nothing happened.

No incident. No damage. No loss.

From an operational perspective, it was success.

From a financial perspective, it was unclear.

"What did we prevent?" someone asked.

There was no number.
No measurable return.
No line item that captured what had been avoided.

Only absence.

And absence, in financial terms, is difficult to defend.

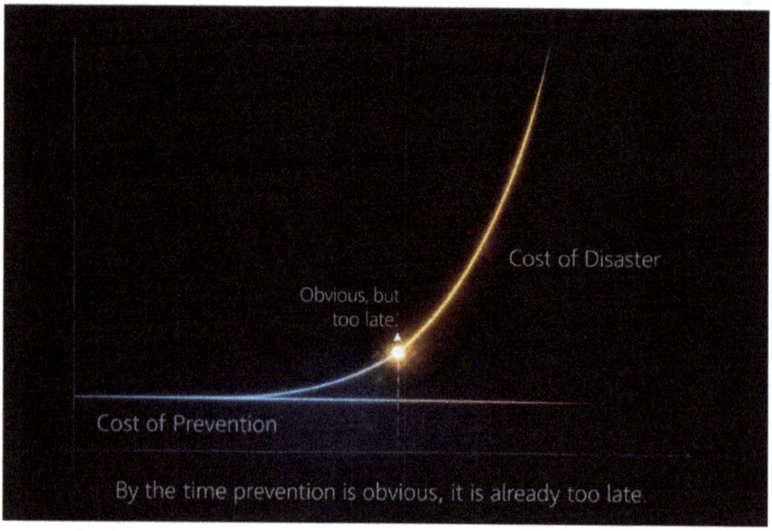

Figure 4.3 Prevention is invisible-until disaster sets its price

Why Prevention Rarely Survives Budget Scrutiny?

Institutions are not irrational.

They allocate capital according to the signals their financial systems are designed to recognise.

Those systems are excellent at funding:

- response
- recovery
- visible damage
- regulatory penalties
- insurance premiums after loss

They struggle to fund:

- foresight
- patience
- resilience
- capacity that may never be "used"

The watcher realises that trust alone is not enough.

Trust authorises action.

Capital sustains it.

Without a financial logic that can carry prevention forward after the moment passes, anticipation collapses back into pilots, proofs, and temporary enthusiasm.

This is not resistance.

It is accounting doing exactly what it was designed to do.

A Familiar Scene: The Budget Room

A budget review is convened.

The early-action system is on the agenda–not as a crisis response tool, but as an ongoing capability. The room is calm. Professional. Curious.

Someone asks how often the system has triggered intervention.

Another asks how much loss has been avoided.

A third asks how this compares to funding response capacity instead.

No one is hostile. No one is dismissive.

But the questions reveal something fundamental:

The institution still values prevention using the language of reaction.

This mismatch is subtle–but lethal to scale.

Chapter 4's Core Claim

Chapter 4 advances a claim that many leaders instinctively feel but rarely articulate:

Prevention fails not because it lacks value, but because its value is measured using the wrong instruments.

You cannot justify foresight using tools designed for hindsight.

> **The Cost That Appeared Too Late**
>
> After the fire, the analysis began.
>
> A post-incident review team reconstructed the timeline.
>
> The signals had been there.
> The patterns had been visible.
> The window for early action had existed.
>
> Now, the cost could be calculated precisely.
>
> Damage. Recovery. Disruption.
>
> The number was clear.
>
> It was far greater than the cost of acting early would have been.
>
> That insight was now obvious.
>
> And entirely too late to matter.

This chapter is about redesigning those tools.

The Shift that Must Occur

For prevention to become durable–not just authorised, but funded–three economic shifts must occur:

- From event-based ROI to trajectory-based ROI
- From cost avoidance to resilience capital
- From pilots to platforms

Each of these shifts is explored in this chapter–not as theory, but as lived institutional experience.

Why Prevention Rarely Gets Funded?

The Problem of Invisible Success

The first-time prevention works, the institution exhales.

The second time, it becomes uneasy.

By the third time, it begins to ask whether the capability is still necessary.

This is the quiet paradox that sits at the centre of prevention economics: success erodes its own evidence.

The watcher has seen this before–not in failure, but in competence. When systems perform well, when nothing escalates, when the world remains stubbornly ordinary, organisations begin to reassess what they are paying for.

Prevention leaves no scars. Response does.

And institutions, whether public or private, are still structurally oriented toward scars.

Modern financial systems are designed to record damage, not restraint.

They excel at capturing:

- losses after failure
- emergency procurement
- overtime and surge staffing
- remediation contracts
- litigation and compliance costs
- insurance claims and premium spikes

These are tangible, auditable, defensible numbers.

Prevention produces none of them.

When early action succeeds, the financial ledger remains quiet. There is no invoice labelled "fire that did not occur." There is no depreciation entry for "system that never failed."

The watcher understands the implication before anyone says it aloud: If value cannot be recorded, then it cannot be defended.

Prevention does not fail in operations. It fails in accounting.

Many prevention programs attempt to survive by quantifying avoided loss.

On the surface, this seems reasonable. If a disaster might have cost millions, then preventing it should be worth millions.

In practice, this logic collapses.

Avoided loss depends on counterfactuals. Counterfactuals invite argument. Argument erodes authority.

The watcher has heard the same sequence unfold repeatedly:

"How certain are we that this would have escalated?"

"Could existing controls have handled it?"

"Was this season unusually mild?"

"Are we attributing too much to the system?"

None of these questions are hostile. All of them are fatal.

The burden of proof placed on prevention is far higher than the burden placed on response. Response only needs a visible event. Prevention must prove an invisible future.

That is an unwinnable contest.

Case Study 1 —

Insurance: When Stability Becomes Suspicion

An insurance group reviews its performance following multiple consecutive seasons without major catastrophic loss.

Claims are down. Combined ratios look healthy. From a distance, the year appears unremarkable.

Inside the organisation, however, something different has occurred.

Early-warning systems flagged persistent environmental stress well ahead of peak season. Underwriting adjusted exposure gradually. Coverage terms were refined quietly. Reinsurance structures were modified to absorb volatility rather than chase yield.

Nothing dramatic happened.

At the executive review, a question surfaces–not from risk, but from finance:

"If risk was increasing, then why didn't we see it in claims?"

The prevention system has done its job too well.

From an accounting perspective, the organisation appears overly cautious. Capital that could have been deployed elsewhere was held back. Returns look conservative. Analysts begin to ask whether underwriting has become too risk-averse.

The watcher sees the irony immediately.

The system that stabilised exposure has undermined its own justification.

Insurance economics reward volatility management only after volatility materialises. Quiet stability reads as missed opportunity, not disciplined foresight.

In this model, prevention looks like under-performance.

Figure 4.4 Why successful prevention suppresses the loss and volatility signals that event-based financial systems are designed to reward, making disciplined foresight appear excessive rather than effective.

The most uncomfortable moment for the insurance organisation does not arrive after loss. It arrives after several quiet years.

The watcher notices how language begins to shift inside the firm. Not abruptly. Not antagonistically. But incrementally, in ways that feel reasonable in isolation.

"Benign conditions."

"Favourable loss experience."

"Conservative posture."

None of these phrases are wrong. Together, they begin to reframe prevention not as discipline, but as restraint bordering on timidity.

In underwriting reviews, the models tell a complicated story.

Exposure indicators have not improved. In some regions, they have worsened. Environmental stressors persist. Infrastructure fragility remains elevated. Nothing in the external risk landscape suggests complacency.

And yet claims are low.

The watcher hears the question that inevitably follows:

"If the risk is rising, then why aren't we seeing it?"

This question is not naïve. It is a structural tension at the heart of insurance economics.

Underwriting models are designed to price manifested risk. When prevention succeeds, it suppresses the manifestation that validates the model.

The better the system becomes at dampening volatility, the harder it is to prove that volatility existed at all.

The tension escalates when capital efficiency enters the discussion.

Capital allocation committees examine return on equity. They compare portfolios. They assess whether capital tied up in certain risk bands could be deployed elsewhere for higher yield.

Prevention complicates this calculus.

Early exposure reduction flattens the loss curve. That stabilisation lowers capital stress but it also lowers the apparent need for capital.

From a narrow perspective, the organisation looks overcapitalised.

The watcher sees how quickly the conversation drifts:

"Are we being paid to take this level of caution?"

"Could we loosen terms without materially increasing exposure?"

These are rational questions inside traditional insurance logic. They are also the beginning of risk re-accumulation.

Reinsurers provide an external reflection of the same paradox.

In renewal discussions, reinsurers acknowledge the insurer's strong controls and disciplined underwriting. Loss experience supports favourable terms.

At the same time, reinsurers recalibrate pricing based on observed outcomes.

Quiet years produce softer pricing signals.

The watcher recognises the recursive risk:

Prevention reduces loss → loss reduces price → price reduces perceived risk → reduced price pressures discipline.

The system begins to forget why it was cautious.

The Boardroom Inflection

The real test comes not in technical forums, but in the boardroom.

A board member, scanning the papers, asks a question that cuts through the abstractions: "Are we sure we're not leaving money on the table?"

No one answers immediately.

Because the honest answer is complicated.

Yes, in the short term, restraint has opportunity cost. Yes, premiums could have been higher. Yes, returns might look better–until they don't.

The watcher understands the difficulty here. Boards are stewards of capital, not guardians of counterfactual futures. They are accountable for visible performance, not invisible discipline.

This is the moment where many prevention programs quietly lose momentum.

Not through rejection.

Through dilution.

The erosion rarely takes the form of a single decision.

It happens through a sequence of small adjustments:

- Slightly broader underwriting appetite
- Incrementally relaxed exclusions
- Gradual return to "market-competitive" terms
- De-prioritisation of early-warning investments

Each adjustment is defensible.

Collectively, they rebuild the same exposure the prevention system worked to flatten.

The watcher has seen this cycle before.

The organisation does not fail prevention because it doubts the intelligence.

It fails because it cannot financially narrate patience.

When Prevention Almost Gets Cut

The most revealing moment arrives during a cost review.

The early-warning capability–analytics, sensing, coordination–appears on a list of discretionary expenditures. Not because it failed, but because it produced no headline justification.

Someone asks whether the system has proven its value.

The watcher knows what the question really means: Can we remove this and still look responsible?

This is the knife-edge moment for prevention.

The Reframing that Saves It

What prevents the cut is not a catastrophic forecast. It is a reframing.

A senior leader intervenes–not emotionally, but structurally: "This system is not here to predict claims.

It is here to prevent volatility from accumulating unnoticed. If we remove it, then we won't see the risk return immediately. We'll see it all at once."

The room quiets.

This framing shifts the logic from profit optimisation to volatility governance.

Volatility, unlike loss, is something boards understand deeply. It destabilises earnings, shakes confidence, and invites external scrutiny.

The watcher recognises this as the moment where prevention finally finds a language capital respects.

The organisation begins to articulate prevention differently:

- Not as avoiding specific disasters
- But as suppressing tail risk
- Not as reducing claims
- But as smoothing capital shock

This reframing aligns prevention with the core mandate of insurance: managing uncertainty over time.

The system's value is no longer tied to whether a particular fire or flood occurred.

It is tied to whether the organisation avoided sudden, destabilising exposure.

The Emerging Economic Logic (unnamed, but present)

Although no one uses formal frameworks yet, the elements of what you later define as L-ARC™ are already visible:

- Anticipatory intelligence preserved time and optionality
- Governance legitimacy allowed early restraint without panic
- Capital sustainability benefited from suppressed volatility

Prevention did not generate profit.

It protected the conditions under which profit remains sustainable.

This is a subtle but decisive shift.

The Final Insurance Lesson

The watcher distils the lesson clearly: Insurance organisations are not punished for missing one disaster. They are punished for being surprised.

Prevention does not exist to eliminate surprise entirely. It exists to ensure that when risk arrives, it arrives gradually–within the organisation's capacity to absorb it.

That capacity has economic value. But only if institutions learn how to recognise it.

Case Study 2 —
Utilities: When Continuity Masks Fragility

The utility does not fail loudly. That is its defining feature.

When it works, it is invisible. When it fails, it becomes unforgettable.

The watcher knows this distinction matters, because utilities are not funded to avoid inconvenience. They are funded to avoid collapse. Everything else–comfort, efficiency, optimisation–comes second.

And yet, the economics governing utilities quietly undermine the very behaviours that prevent collapse.

In the control room, the screens are calm.

Load flows are balanced. Frequencies stay within tolerance. Automated controls compensate for small deviations without escalating alarms. The system is doing exactly what it was designed to do.

What the watcher sees, however, is not calm. It is strain without rupture.

Assets are operating closer to limits more often. Maintenance windows shrink. Environmental conditions behave differently than historical baselines assumed. Nothing breaks–but nothing fully recovers either.

This is not instability. It is erosion.

Advanced monitoring systems begin to surface patterns that traditional outage metrics miss.

Equipment degradation accelerates subtly

Thermal margins thin more frequently

Weather-induced variability persists longer than expected

Manual intervention becomes routine rather than exceptional

Each signal alone is manageable.

Together, they suggest a system operating with less slack than its designers intended.

The watcher raises the issue cautiously. The language remains technical, restrained, defensible.

The response is predictable:

"As long as we're not seeing outages, we're within acceptable limits."

In utilities, continuity substitutes for legitimacy.

Figure 4.5 How uninterrupted service conceals accumulating system stress, delaying intervention until decision options narrow and emergency capital becomes unavoidable

Utility economics reward efficiency.

Capital expenditure is scrutinised heavily. Regulators pressure operators to do more with less. Consumers expect reliability without visible investment.

Prevention complicates this equilibrium.

Early intervention often looks like:

- accelerated asset replacement
- additional redundancy
- pre-emptive maintenance
- operational derating

All of these reduce short-term efficiency.

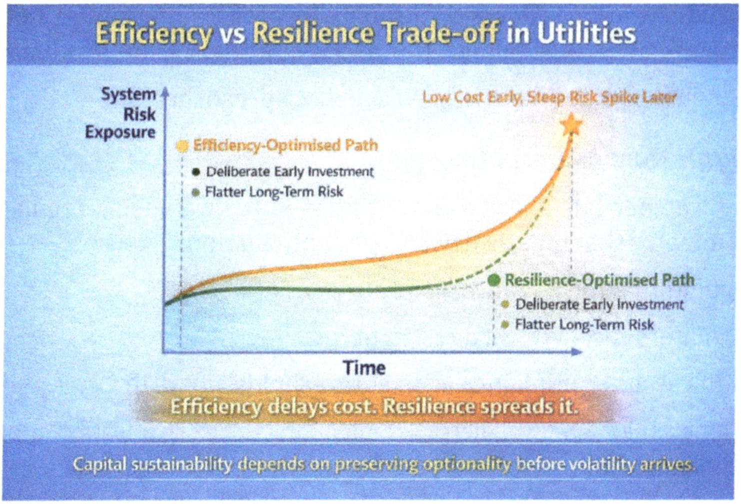

Figure 4.6 How efficiency-optimised systems defer cost until it returns suddenly as emergency capital, while resilience-optimised systems distribute cost earlier to preserve decision optionality.

From a regulatory lens, they can appear excessive.

From a financial lens, they look like cost inflation without incident justification.

The watcher recognises the bind immediately.

If the system fails, then leadership is blamed for negligence.

If the system holds, then leadership is questioned for overspending.

There is no neutral outcome.

The tension becomes explicit during a capital review.

Engineering presents a case for early reinforcement of specific network segments. The justification is careful: not imminent failure, but persistent stress and compounding exposure.

Finance listens patiently.

Then the familiar question arrives:

"What outage does this prevent?"

The engineer hesitates.

Because the honest answer is: none in particular.

The intervention reduces the probability of cascading failure, not the certainty of a single outage. It preserves resilience, not headlines.

That answer does not travel well across spreadsheets.

Regulatory Asymmetry

The watcher has learned that utilities operate under a regulatory asymmetry that quietly discourages anticipation.

Regulators punish failure visibly

Regulators rarely reward restraint invisibly

Audits assess compliance against thresholds, not drift

As long as service-level metrics remain intact, the organisation appears compliant.

Compliance becomes the proxy for safety.

This is a dangerous substitution.

Eventually, conditions converge.

Not dramatically. Not catastrophically.

But enough that the organisation considers a proactive network reconfiguration–temporary load reduction, pre-emptive derating, selective service limitation.

The proposal is modest. Reversible. Carefully scoped.

It is also unprecedented.

The watcher senses the hesitation before it is voiced.

"What if we act and nothing happens?"

This question carries weight in utilities.

Acting without visible failure invites public scrutiny. It raises questions about necessity, transparency, and justification. It risks being perceived as overreach.

Waiting feels safer.

The Silent Cost of Waiting

Waiting works until it doesn't.

But before that moment arrives, it exacts quieter costs.

Maintenance becomes reactive.

Operator fatigue increases.

System margins narrow.

Emergency flexibility declines.

The network continues to operate, but its ability to absorb surprise diminishes.

This erosion does not appear in outage statistics. It appears in decision optionality.

Ironically, the first external signal does not come from operations. It comes from insurers.

Premiums adjust subtly. Terms tighten. Questions increase. Insurers do not see failure, but they see coherence or the lack of it.

Markets price trajectory, not reassurance.

The watcher understands the warning immediately.

When insurers hesitate, capital is already listening.

The Reframing that Changes the Conversation

What finally shifts the debate is not an outage.

It is a reframing of purpose.

A senior leader intervenes:

"We are not investing to prevent outages. We are investing to prevent loss of control."

This language resonates.

Loss of control is something utilities fear more than downtime. It implies cascading effects, regulatory intervention, political fallout.

Prevention is no longer framed as avoiding failure.

It is framed as preserving governability.

The Economic Realisation

Once framed this way, the economics begin to align.

Early intervention:

- reduces emergency capital spikes
- smooths regulatory exposure
- stabilises insurance relationships
- protects long-term asset value

These benefits do not improve quarterly efficiency.

They protect institutional survival.

The watcher sees the shift happen quietly.

Prevention is no longer justified by hypothetical disasters. It is justified by the cost of being cornered.

The Unspoken Framework Emerging

Again, without naming it, the structure becomes visible:

Anticipatory intelligence identifies stress early.

Governance legitimacy allows action without crisis.

Capital sustainability benefits from smoother exposure.

The utility has not become braver. It has become structurally prepared.

The Utility Lesson

Utilities do not fail because they ignore risk. They fail because they mistake continuity for safety.

Prevention disrupts that illusion.

But only when institutions learn to value fragility reduction, not just uptime.

Bridge to the Public Sector Case

The watcher recognises that utilities occupy a middle ground.

Less exposed than insurance to market volatility.

Less protected than government by political mandate.

If prevention struggles here–where consequences are tangible and systems are engineered–then the challenge will be even sharper in the public sector.

That is where the story must go next.

Case Study 3 —

When the Public Sector Pays for What It Did Not Fund (Public Finance, Treasury, and the Cost of Invisible Prevention)

The third case does not begin with an incident.

It begins with a budget cycle.

On paper, the system looks stable. Expenditure is within forecast. Service levels are intact. There is no emergency appropriation request, no cabinet briefing marked urgent, no line item screaming for attention. From a fiscal standpoint, the year is unremarkable–which, in the public sector, is often treated as success.

Inside the department, however, a quieter conversation has been unfolding.

It does not use the language of crisis. It uses the language of constraint.

There are projects that cannot quite proceed. Pilots that remain pilots longer than expected. Maintenance schedules that stretch, then stretch again. Capital refresh programs that are re-sequenced–not cancelled, but delayed just enough to fit within existing envelopes.

Each decision is defensible. Each one is incremental. Each one is framed as prudent stewardship.

This is how public systems adapt.

And this is where the cost begins to accumulate.

The Illusion of Fiscal Discipline

In the public sector, prevention faces a unique burden.

Unlike insurance, there is no premium signal.

Unlike utilities, there is no immediate service interruption.

Instead, legitimacy is tied to visible restraint.

A department that spends less than forecast is praised.

A minister who avoids new funding requests is seen as disciplined.

A treasury that resists speculative investment is viewed as responsible.

Prevention, by contrast, looks like anticipation without proof.

It asks for money when nothing has yet gone wrong.

The watcher in this system is not an operator or an engineer.

They are a policy analyst.

Their work is upstream–buried in modelling assumptions, inter-agency briefings, and long-range risk scenarios that are always framed as contingencies, never certainties. They see the same patterns as before: persistence, convergence, drift.

But here, the challenge is sharper.

Because in public finance, spending early is not neutral.

It competes directly with hospitals, schools, and visible voter priorities. Every dollar spent preventing something abstract must be defended against something tangible.

So, the language shifts.

Risk becomes "manageable."

Exposure becomes "within tolerance."

Prevention becomes "nice to have."

Nothing is denied outright.

Nothing is approved either.

Budgeting for Events, Not for Absence

The capital framework is the problem.

Public budgeting is event-oriented by design. Funding is released when something happens–a disaster, a failure, a breach that can be named, audited, and explained. Supplementary appropriations follow consequence, not anticipation.

This creates a structural asymmetry:

Money flows easily after harm

Money struggles to flow before harm

The watcher knows this. They have seen it play out before.

After floods.

After fires.

After infrastructure failures that suddenly justify emergency capital at ten times the cost of earlier intervention.

They also know something else.

Once emergency funding arrives, it arrives without optionality.

Contracts are rushed.

Procurement shortcuts are taken.

Governance tolerances are temporarily suspended.

The system acts–but it acts late, loudly, and expensively.

This is not inefficiency.

It is inevitability baked into the funding model.

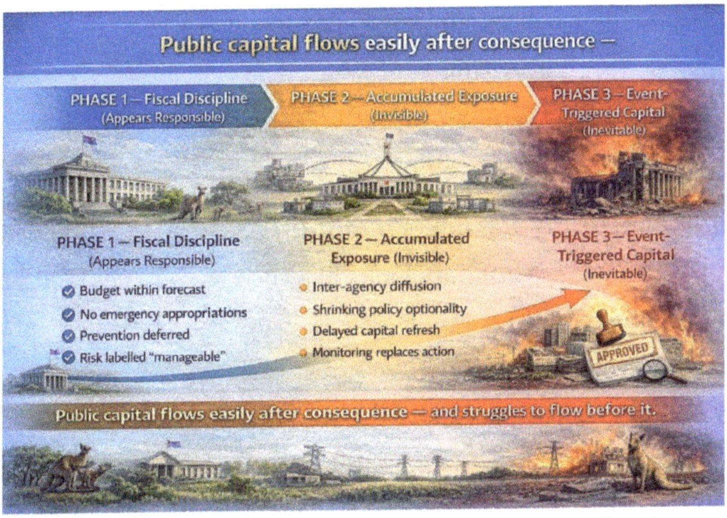

Figure 4.7 How public-sector funding models prioritise fiscal restraint early, while compounding risk accumulates invisibly until emergency capital becomes unavoidable.

A Near-Intervention That Fails Quietly

There is a moment–there is always a moment.

A proposal circulates across departments. It is careful. Conservative. It does not predict catastrophe. It argues for modest, staged investment to preserve flexibility across scenarios that are becoming more likely, not more certain.

The proposal survives initial review.

It does not survive prioritisation.

The reason is never written down as rejection. Instead, it is deferred into the future budget cycle, where it must compete again–not against risk, but against immediacy.

The watcher reads the feedback closely.

The concern is not that the analysis is wrong.

The concern is that the absence of an event makes the investment politically fragile.

This is the public-sector version of the same pattern you have already seen:

- intelligence is accurate
- patterns are visible
- legitimacy is missing

And without legitimacy, capital waits.

When Continuity Becomes a Liability

From the outside, continuity looks like success. Services continue.

Communities adapt.

Temporary measures become routine.

Internally, however, continuity is masking fragility.

Risk is not increasing in dramatic steps. It is spreading laterally across agencies, jurisdictions, and funding lines. What once sat inside a single department now touches multiple portfolios, none of which fully own the problem.

This diffusion of ownership makes early action harder, not easier.

No single minister wants to be the first mover.

No single agency can justify the spend alone.

No single budget line captures the benefit.

Capital becomes trapped between silos.

And so, the system waits.

The Cost That Arrives Without Warning

When the moment finally arrives, it does not feel sudden to those who have been watching.

But it arrives suddenly to everyone else.

A failure.

A compounding event.

A public inquiry that asks why action was not taken earlier.

Emergency funding is released quickly now. There is no debate about legitimacy anymore. The evidence is visible. The consequences are real.

But the options are gone.

What could have been staged is now urgent.

What could have been modular is now monolithic.

What could have been adjusted is now locked in.

The public sector pays–not just financially, but institutionally.

Trust erodes.

Scrutiny intensifies.

Future funding becomes more constrained, not less.

This is the paradox: late spending damages future spending capacity.

Reframing Public Capital Through Lahiri AI Resilience Continuum (L-ARC™)

This is where L-ARC™ matters most.

Not as a technical framework. Not as an AI story.

But as a capital legitimacy architecture.

L-ARC™ reframes public investment away from predicting events and toward preserving decision optionality under uncertainty.

It offers public leaders a different justification:

- We did not fund certainty.
- We funded flexibility.

- We did not prevent a disaster.
- We prevented the loss of choice.

In the public sector, this distinction is everything.

Because legitimacy does not come from being right early. It comes from being explainable later.

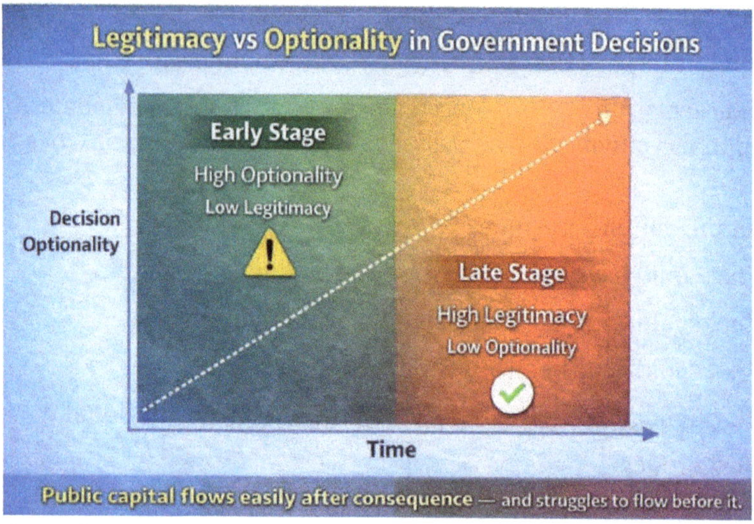

Figure 4.8 How public-sector decision legitimacy typically increases only as decision optionality declines

L-ARC™ provides the missing bridge:

- between intelligence and authority
- between anticipation and audit
- between early action and political defensibility

The Practitioner Takeaway (Public Sector)

For public leaders, the lesson is not that prevention is cheap.

It is that waiting is structurally expensive, even when it looks fiscally conservative.

Capital sustainability in government depends on:

- Recognising that continuity can conceal fragility.
- Accepting that legitimacy must be designed, not awaited.

- Funding the preservation of options, not just the repair of damage.

This is not about acting on fear.

It is about acting while choice still exists.

By this point in the story, the failure of prevention is no longer mysterious.

The signals were present. The models were accurate. The exposure was visible–at least to those trained to see it.

Patterns persisted where randomness was expected.

Anomalies clustered where variance once explained them away.

Near-interventions surfaced, quietly and repeatedly, before being folded back into dashboards and monthly reports.

From a technical perspective, the system worked.

And yet, no capital moved.

This is the moment where most post-mortems reach for comfortable explanations: denial, political hesitation, risk aversion, short-termism. These explanations feel intuitive and they are often wrong.

The failure here is not one of belief.

Institutions did not disbelieve the risk.

They simply could not authorise it.

What stalled prevention was not ignorance or incompetence.

It was translation.

Risk arrived early–but it arrived speaking the wrong language.

Institutions do not reject prevention because they question the data. In fact, senior leaders increasingly accept that modern risk is complex, non-linear, and difficult to forecast precisely. What they cannot accept is action that cannot be justified, defended, and sustained under scrutiny.

Risk may be technically compelling, but capital is procedurally conservative.

Finance committees are not designed to reward foresight. They are designed to protect legitimacy.

They do not approve spending because something might happen.

They approve spending when action can be explained clearly, audited later, and defended publicly–even if nothing happens at all.

This distinction matters more than most technical teams realise.

Prevention fails not because it is speculative, but because it lacks a narrative of legitimacy that survives success. When nothing happens–as prevention intends–there is no visible event to validate the decision. No incident to point to. No aftermath to justify the cost. Silence becomes ambiguity, and ambiguity becomes vulnerability.

In event-based systems of accountability, quiet outcomes look indistinguishable from unnecessary ones.

As a result, prevention is judged not on what it avoids, but on what it fails to visibly produce.

Until anticipatory action can meet the institutional standard of explainability, defensibility, and endurance–until it can justify itself in the absence of harm–it will always appear excessive. Over-engineered. Premature. Politically risky.

This is where most technical systems fail.

Not in sensing.

Not in modelling.

Not in prediction.

They fail at the precise point where intelligence must become permission.

Figure 4.9 - How anticipatory intelligence becomes investable only when governance legitimacy and capital design are established before crisis, allowing prevention to succeed without requiring post-event justification

Detection produces awareness.

But decision conversion requires legitimacy.

And without that conversion layer, even the most accurate intelligence will continue to wait–quietly, correctly, and ineffectively–until consequence finally forces capital to move.

What the system sees and what capital requires are structurally different:

Systems observe patterns.

Institutions authorise decisions.

Capital demands defensibility.

Prevention fails when these layers remain misaligned.

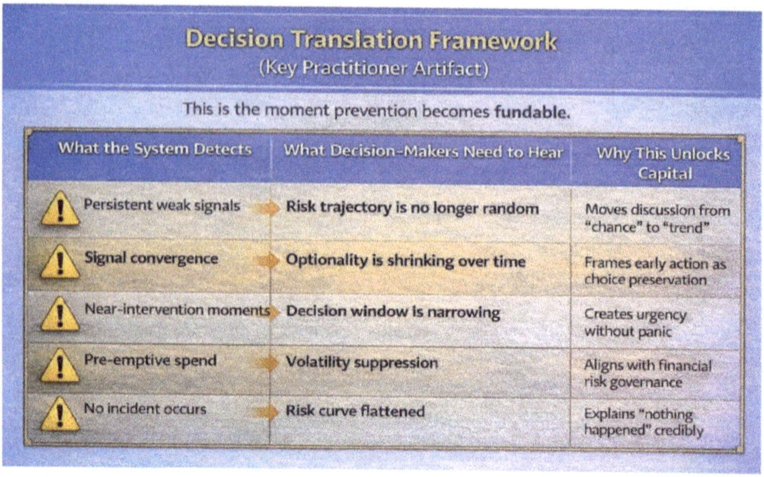

Decision Translation Framework
(Key Practitioner Artifact)

This is the moment prevention becomes **fundable.**

What the System Detects	What Decision-Makers Need to Hear	Why This Unlocks Capital
Persistent weak signals	**Risk trajectory is no longer random**	Moves discussion from "chance" to "trend"
Signal convergence	**Optionality is shrinking over time**	Frames early action as choice preservation
Near-intervention moments	**Decision window is narrowing**	Creates urgency without panic
Pre-emptive spend	**Volatility suppression**	Aligns with financial risk governance
No incident occurs	**Risk curve flattened**	Explains "nothing happened" credibly

Fig 4.10 What institutions require at this point is not more accuracy, but a translation mechanism that converts signals into legitimate, fundable decisions.

This translation does not exaggerate risk. It legitimises action.

Across insurance, utilities, and public-sector finance, the same approval conditions appear again and again:

Evidence of persistence, not prediction certainty.

A clear governance chain authorising early action.

Financial framing tied to volatility control, not loss avoidance.

Audit-ready explanation for why action occurred before harm.

Capital designed to remain justified even if outcomes stay quiet.

Prevention succeeds not when it proves danger–but when it proves prudence.

This is the structural role L-ARC™ begins to play.

Not as a technology.

Not as a model.

But as a decision architecture that ensures:

Intelligence arrives early. Legitimacy arrives before crisis.

Capital arrives while optionality still exists

When those three move together, prevention stops looking excessive–and starts looking inevitable.

Prevention does not fail because it is invisible. It fails because institutions have not yet learned how to approve success that leaves no scar.

At this point, the L-ARC™ continuum completes its arc.

Anticipatory intelligence has learned how to speak.

Governance has learned how to listen.

Capital, for the first time, has learned how to wait.

Valuing What Didn't Happen

ROI, Avoided Loss, and Resilience Capital

By the time the organisation begins asking how to value prevention, something subtle but irreversible has already occurred.

It has stopped asking whether risk exists.

That question was settled quietly, weeks ago.

What remains unsettled is whether acknowledging that risk is enough to justify acting on it.

Ecological stability functions as unpriced capital. Stable habitats reduce emergency response costs, preserve land usability, protect tourism and agricultural confidence, and lower long-term public liability. When AI-enabled prevention stabilises ecosystems, it does not merely avoid loss–it preserves future economic optionality that traditional ROI models fail to capture.

The watcher notices the shift first–not on screens this time, but in meetings. The tone has changed. The language has softened. People no longer dispute the presence of exposure; they debate its meaning. They debate proportionality. They debate justification.

The conversation has moved from perception to permission.

This is where most prevention initiatives stall–not because leaders doubt intelligence, but because they cannot yet price restraint.

Prevention has an accounting problem.

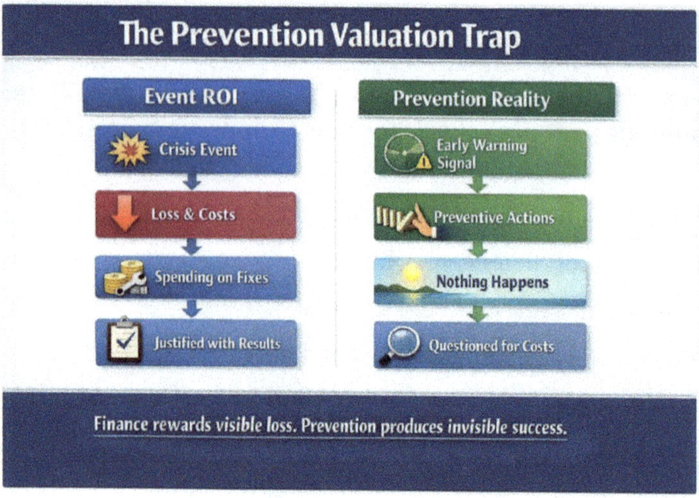

Figure 4.11 Finance rewards visible loss. Prevention produces invisible success

When response succeeds, damage is visible. Loss is measurable. Spend is justified retroactively by what everyone can see.

When prevention succeeds, nothing happens. No smoke.

No outage.

No evacuation.

No headline.

The organisation continues operating, and paradoxically, that continuity becomes evidence that the investment was unnecessary.

Finance teams are not hostile to prevention. They are allergic to non-events. A senior finance leader says it plainly in one closed session:

"If nothing happens, then how do we explain why we spent this money?"

No one challenges the logic.

It is not cynical.

It is fiduciary.

Capital must be defended not just at approval, but in hindsight–especially when outcomes are quiet.

Figure 4.12 Institutions fail not from losses, but from the erosion of resilience capital

This is the core dilemma: prevention must justify itself in a world that only rewards visible failure.

Case Layer 1: Insurance as a Forward-Looking Witness

The first crack in this logic comes not from operations or technology, but from outside the organisation.

It comes from insurers.

Insurers do not wait for certainty. They respond to coherence.

They do not ask whether a fire will occur tomorrow; they ask whether the system appears predictable enough to remain insurable.

In one review cycle, the organisation notices a small change.

Premiums have not surged. Coverage has not been withdrawn.

But terms have tightened.

Exclusions are clarified.

Deductibles inch upward.

Language around "foreseeable risk" appears more frequently.

No one panics. This is not a crisis.

But it is a signal.

Insurance pricing is one of the earliest places where prevention quietly shows its value–not as savings, but as optionality.

When prevention systems are credible, insurers do not reward them with applause.

They reward them with patience.

Patience shows up as slower premium growth, broader coverage, fewer exclusions, and a longer horizon before risk becomes uninsurable.

These are not line items most boards track.

But they are forms of capital preservation.

The watcher recognises this immediately. The organisation does not–yet.

The first failed ROI Conversation

A working group is formed to "build the business case."

It does what organisations always do when facing ambiguity.

It reaches for historical loss data.

Spreadsheets appear.

Charts are generated.

Past incidents are modelled.

The numbers are sobering–but incomplete.

Because historical loss only measures what broke, not what almost did.

The group struggles.

Every model assumes a counterfactual that cannot be proven:

What would have happened, if we had not acted?

This is where traditional ROI frameworks fail prevention.

They treat avoided loss as speculative, not structural.

They ask prevention to meet the same evidentiary standard as response, despite operating on a different temporal logic.

Prevention works upstream of proof.

Finance knows this.

Operations feels this.

Governance struggles to reconcile it.

The meeting ends with no decision–not rejection, but deferral.

Again.

The breakthrough does not arrive as a spreadsheet. It arrives as a reframing.

A senior leader–quiet, experienced–asks a different question: "What are we actually trying to protect here?"

The room pauses.

Not assets.

Not infrastructure.

Not even people, directly.

What prevention protects first is optionality.

The ability to choose how to respond, rather than being forced to respond at all.

That insight changes the valuation problem.

Because optionality can be priced–just not through loss models.

The organisation begins to recognise a form of value it has never formally named. Not revenue.

Not savings. Not insurance offsets.

But resilience capacity.

Resilience capital shows up as:

- Time gained before escalation
- Fewer forced decisions
- Lower volatility in planning
- Reduced dependence on emergency appropriations
- Greater credibility with regulators and insurers
- More stable workforce conditions
- Strategic freedom under stress

None of these appear in a single ledger.

Together, they determine whether an organisation governs risk–or is governed by it.

This is the missing layer between intelligence and action.

And it is precisely the layer most ROI models ignore.

Case Layer 2: Utilities and the Price of Latency

In utilities, the economics sharpen.

Here, prevention is not abstract–it competes directly with reliability metrics.

The organisation reviews a near-miss involving infrastructure stress under extreme conditions.

Nothing failed.

Nothing ignited.

But response readiness consumed extraordinary resources.

Crews were redeployed.

Maintenance was deferred.

Contingency budgets were drawn down.

No damage occurred. But costs did.

The watcher notices the pattern: the absence of failure does not mean the absence of expense.

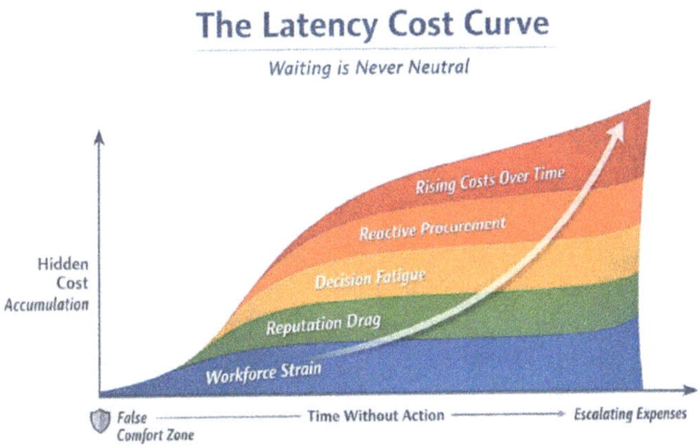

Figure 4.13 The Latency Cost Curve: How waiting converts invisible strain into escalating expense, even when no incident occurs

Latency has a price.

The longer the organisation waits to act on early signals, the more it pays–not in losses, but in friction.

Overtime.

Emergency procurement.

Reputational exposure.

Decision fatigue.

These are real costs.

They are simply unlabelled.

When reframed this way, prevention no longer competes with response.

It competes with chronic inefficiency.

A revised proposal returns–this time grounded differently.

Not in catastrophic scenarios.

Not in fear.

But in volatility reduction.

The proposal does not promise to prevent fires.

It promises to:

- Reduce operational variability
- Stabilise planning horizons
- Improve insurer confidence
- Lower emergency cost spikes
- Preserve executive decision space

This language resonates.

Because it speaks finance.

The proposal still does not pass unanimously.

But this time, it is not deferred. It is conditionally approved.

With guardrails.

With milestones.

With governance oversight.

Prevention has finally learned how to speak.

The organisation learns something crucial in this moment:

Prevention is not funded when risk is convincing. It is funded when action is defensible even in hindsight.

This is the bridge this section builds.

Between intelligence and capital.

Between foresight and permission.

Between knowing and acting.

What emerges is a new valuation logic–one that does not ask: "What disaster did we prevent?"

But instead:

"What instability did we avoid introducing into the system?"

That is a question boards can answer.

That is a question capital understands.

By the end of this section, the organisation no longer sees prevention as an expense waiting for justification.

It sees it as infrastructure for choice.

The language has changed.

The framing has shifted.

The economics have matured.

What remains unresolved is scale.

Pilot economics are persuasive.

Enterprise economics are harder.

Because prevention that cannot scale becomes another well-intentioned experiment.

And intelligence that cannot scale becomes just another warning that arrived early–and went unheard.

The conditional approval does not feel like a victory.

It feels provisional.

Prevention has been allowed to exist, but only under supervision. Only within limits. Only as long as it behaves.

This is how institutions domesticate unfamiliar value.

They do not reject it outright. They place it inside governance structures designed for something else and watch to see whether it survives.

The watcher understands this instinctively. They have seen it before–new ideas granted permission only if they agree to remain small.

Small enough not to disrupt budgeting norms.

Small enough not to threaten precedent.

Small enough to be reversible.

Reversibility is the quiet requirement no one names.

The first review arrives before the system has had time to prove anything.

Not because it has failed–but because it has succeeded too quietly.

No incident has occurred.

No emergency has been averted publicly.

No metric has moved dramatically.

This is exactly the outcome prevention aims for.

It is also exactly what makes it vulnerable.

The review focuses on utilisation.

"Are we using this enough?"

"Is this proportionate to the spend?"

"Could we have waited longer?"

These are reasonable questions.

They are also downstream questions–asked after the decision window has already narrowed.

The watcher notices how often the word enough appears.

Enough signals.

Enough certainty.

Enough justification.

Enough, in this context, means obvious.

Prevention will never be obvious.

Case Layer 3: Public Sector Economics and the Tyranny of Optics

In the public sector, the valuation problem intensifies.

Here, prevention competes not just with other priorities, but with visibility itself.

A transport agency reviews its risk posture after a quiet season.

No major failures.

No public disruption.

No political fallout.

Treasury asks the predictable question: "If nothing happened, then why do we need to increase funding?"

The agency has data.

It has models. It has convergence analysis.

What it lacks is a narrative that survives inquiry.

Public capital is accountable not only to outcomes, but to perception. Spending that prevents disaster but produces no story is always vulnerable.

In this context, prevention is not merely unfunded–it is politically expensive.

The organisation learns something uncomfortable: The quieter the success, the harder it is to defend.

Why Traditional ROI Collapses under Scrutiny?

Another attempt is made to formalise ROI.

This time, the model is more sophisticated.

It includes:

- Probabilistic loss curves
- Scenario-weighted outcomes
- Sensitivity analysis

The model is technically sound.

It still fails.

Because ROI framed as avoided catastrophe invites a single, devastating question: "What if the catastrophe was never going to happen?"

No model can answer that conclusively.

The watcher realises that prevention cannot win if it insists on arguing counterfactuals. Counterfactuals invite scepticism.

What prevention needs is structural evidence, not hypothetical proof.

The turning point comes unexpectedly–from comparison rather than prediction.

The organisation begins benchmarking itself against peers.

Not on incidents. On volatility.

One region's operations show tighter variance.

Another's emergency expenditure spikes repeatedly.

A third shows growing insurance friction.

Patterns emerge–not of failure, but of strain.

The watcher notices how these comparisons change the conversation.

No one asks: "Why did nothing happen here?"

They ask: "Why is it calmer here than elsewhere?"

Calm becomes evidence.

Stability becomes performance.

This is subtle, but profound.

Prevention no longer has to prove that disaster was imminent.

It only has to demonstrate that its absence of volatility is not accidental.

Reframing Success as Comparative Advantage

The organisation begins speaking differently.

Not about fires avoided. But about options preserved.

Not about money saved. But about costs not forced.

Not about predictions. But about decision space.

A new phrase enters the lexicon: "Operational headroom."

Headroom is something executives understand. It explains why:

- Decisions feel less urgent
- Planning horizons feel longer
- Stress feels lower but not complacent

Headroom does not eliminate risk. It makes risk governable.

Case Layer 4: Insurance as a Delayed Validator

The next insurance renewal confirms what the organisation cannot yet prove internally.

While premiums rise across the sector, theirs rise more slowly.

While exclusions proliferate elsewhere, theirs remain narrow.

No one congratulates them.

No one attributes this explicitly to prevention.

But capital markets are rarely explicit.

They signal through behaviour.

Insurance has become a proxy for legitimacy.

Not because insurers trust prevention models.

But because they trust patterns of restraint.

The watcher recognises the irony:

The organisation still struggles to justify prevention internally.

Externally, it is already being priced as lower risk.

When Audit Becomes an Ally

An audit review, expected to challenge the initiative, produces a different outcome.

The auditors do not question the absence of incidents. They question the consistency of decision-making.

They notice:

- clearer escalation logic
- documented early rationale
- fewer ad-hoc responses

For the first time, prevention shows up as governance maturity. Not as spend.

Not as technology.

As process coherence.

This is the moment the organisation realises something essential:

Prevention does not need to be justified as foresight. It can be justified as discipline.

The Economics of Discipline

Discipline has a price–but also a return.

It reduces:

- emergency premiums
- reactive procurement

- political exposure
- decision fatigue

None of these are traditionally attributed to prevention.

They should be.

The watcher sees the ledger differently now.

Not as a balance of costs and savings, but as a balance of forced decisions versus chosen ones.

Forced decisions are always expensive.

Chosen decisions can be optimised.

This is the economic heart of prevention.

By the second review cycle, something has changed.

The questions are different.

No one asks: "Was this necessary?"

They ask: "How do we ensure this remains proportionate?"

That shift matters.

Necessity is binary.

Proportion is adjustable.

Prevention has moved from being justified to being managed.

This is legitimacy.

Why Scaling Remains Hard?

Despite progress, resistance remains.

Scaling prevention means:

- embedding it into budgets
- normalising early spend
- accepting quiet success

Each of these threatens existing reward structures.

Response heroes are visible.

Prevention architects are not.

The organisation understands now that scaling prevention is not a technical problem. It is a cultural one.

And culture moves slower than capital.

The real test of prevention does not come when it is approved.

It comes when it must be defended repeatedly, quietly, and without drama.

Once the initial permission is granted, the organisation enters a more difficult phase–one that receives far less attention but determines whether prevention becomes structural or collapses back into exception.

This is the phase where prevention must coexist with normal operations. And normal operations have long memories.

Institutions remember crises vividly.

They remember dates, decisions, and faces. They remember the urgency, the visibility, the authority that comes with acting under pressure. Those memories reinforce the belief that meaningful action requires dramatic context.

What institutions do not remember well are periods of restraint.

Quiet seasons blur together. Stable years collapse into a single, undifferentiated past. The absence of catastrophe does not lodge itself in organisational memory as achievement.

This creates an asymmetry.

Response is remembered as leadership.

Prevention is remembered–if at all–as overhead.

The watcher senses this shift in how conversations subtly reference the past. Crises are named. Quiet periods are not.

Prevention is working against forgetfulness.

As time passes, the organisation begins to ask a dangerous question:

"Have conditions actually improved, or did we overestimate the risk?"

This question does not arise because intelligence has weakened. It arises because the absence of consequence invites reinterpretation.

Humans are pattern-seeking–but institutions are story-seeking.

They prefer narratives with beginnings, climaxes, and resolutions. Prevention denies them resolution.

Instead, it offers continuity.

Continuity feels ambiguous.

Ambiguity invites doubt.

This is why prevention must eventually stop arguing for itself as risk mitigation and begin arguing for itself as institutional posture.

Case Layer 5: Infrastructure Planning and Deferred Temptation

In a long-term infrastructure planning session, the tension surfaces explicitly.

Capital budgets are under pressure. New projects compete for attention. A proposal emerges to defer a prevention-related upgrade–just for one cycle.

Nothing has gone wrong.

Systems are stable.

Metrics are green.

The deferral is framed as pragmatic.

"We can afford to wait one more year."

No one disputes the logic.

The watcher notices something else.

The argument for deferral relies entirely on the success of prevention to date. The very stability created by early investment is now being used as justification to unwind it.

This is the paradox of effective prevention.

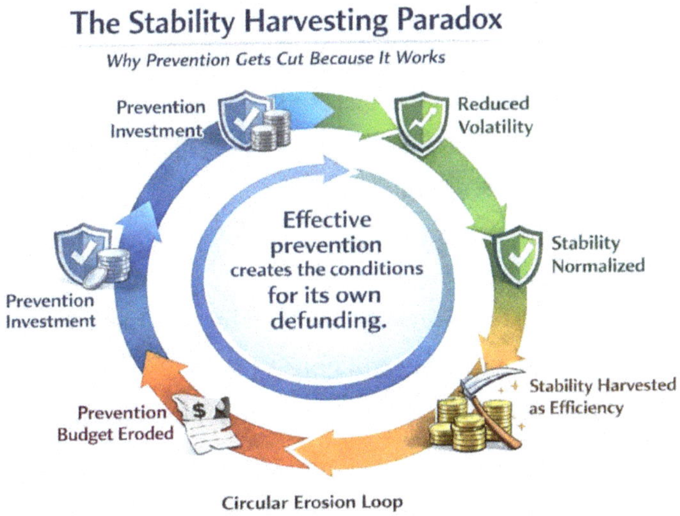

Figure 4.14 Prevention succeeds, volatility falls, budgets shrink–and risk returns unnoticed

It creates the conditions under which it appears unnecessary.

As stability becomes normalised, a new behaviour emerges: extraction.

Departments begin reallocating resources made available by reduced emergency spend. Planning assumptions quietly incorporate the absence of disruption as baseline.

Stability is harvested as efficiency.

Efficiency, once gained, is rarely returned.

The organisation has not reversed prevention–but it has begun to consume its benefits prematurely.

This is not malicious. It is structural.

Institutions are optimised to convert slack into output.

Resilience produces slack. Slack attracts pressure.

The watcher realises that prevention fails not when it is denied, but when it is absorbed too completely into normal budgeting logic.

Once prevention is treated like any other cost centre, it becomes vulnerable to incremental erosion.

Cuts are never dramatic.

They are fractional.

They accumulate.

A capability is weakened without anyone deciding to weaken it.

This is where governance must evolve.

Not to approve prevention once–but to protect it from being silently dismantled.

A subtle but important language shift begins to appear.

Leaders stop asking: "Do we still need this?"

They begin asking: "How do we ensure this remains effective?"

This marks a transition.

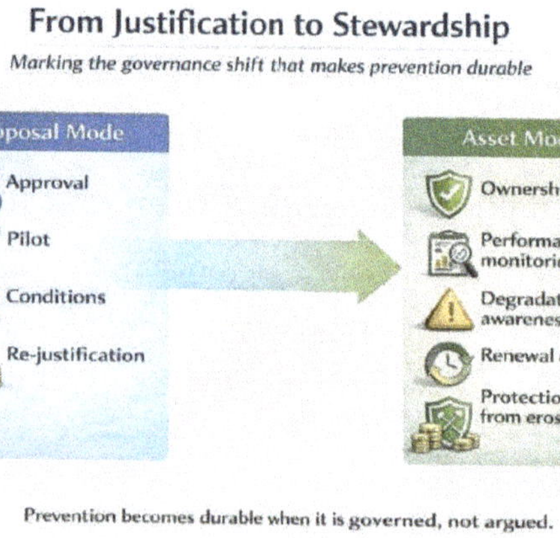

Figure 4.15 This is the moment prevention ceases to be a proposal and becomes institutional responsibility

Prevention is no longer treated as a proposal. It is treated as an asset.

Assets require stewardship.

Stewardship implies:

- ➢ performance monitoring
- ➢ degradation awareness
- ➢ renewal cycles
- ➢ clear ownership

This is not how prevention has traditionally been managed.

But it is how capital-intensive systems survive.

Case Layer 6: Workforce Signals as Economic Indicators

The organisation notices another unexpected effect.

Turnover stabilises.

Absenteeism declines slightly.

Fatigue indicators soften.

None of these shifts are dramatic enough to headline. Together, they matter.

Human systems respond early to sustained pressure. When prevention reduces volatility, people feel it before metrics capture it.

The watcher sees this reflected in informal conversations.

People speak less about being "on edge."

They plan leave further in advance.

They stop bracing for escalation.

This is not morale.

It is operational confidence.

Operational confidence has economic value.

It reduces error.

It improves judgment.

It preserves institutional memory.

None of this appears in ROI models.

All of it affects outcomes.

At this point, the organisation begins to understand prevention as something broader than risk management.

It is a workforce stabilisation mechanism.

Reactive environments burn people out.

Anticipatory environments retain them.

Retention reduces recruitment costs.

Experience reduces error.

Judgment improves under stability.

The economics of prevention extend into human capital–even if accounting frameworks lag behind.

Resistance persists in one familiar form.

"Even if we scale this back, we can always respond if needed."

This belief is comforting and flawed.

Response capacity degrades quietly when not exercised under real pressure. Teams adapt to stability. Muscle memory fades. Coordination assumptions shift.

The organisation realises something counterintuitive:

Sustained prevention requires intentional response rehearsal, not complacency.

Prevention does not replace response. It reshapes it.

This insight further complicates valuation.

Prevention must now justify not only its existence, but its coexistence with readiness.

By this stage, the organisation is no longer trying to squeeze prevention into a single ROI narrative.

It recognises multiple value streams:

- volatility suppression
- decision optionality
- insurance patience
- workforce stability
- governance coherence
- audit confidence

None of these alone justify prevention.

Together, they make the absence of prevention indefensible.

This is the real turning point.

Prevention no longer needs to prove it saves money.

It proves that without it, the organisation would be structurally weaker–even if no disaster occurred.

Once prevention is understood this way, it becomes difficult to unwind without explicit acknowledgement of what would be lost.

Cuts would no longer be neutral. They would be directional.

They would trade:

- calm for urgency
- choice for compulsion
- discipline for reaction

Institutions hesitate to make such trades openly.

This is how prevention becomes durable.

Not through certainty. But through embedded dependence.

The watcher realises something that would have sounded impossible earlier in the story.

Prevention has stopped asking for permission.

It has become part of how the organisation understands responsible leadership.

Not heroic. Not dramatic.

But deliberate.

That is its quiet success.

Yet one challenge remains unresolved.

Everything described so far works because prevention is still partially exceptional. Still supervised. Still explained.

Scaling it–embedding it across systems, regions, agencies–will introduce new risks.

Consistency risk.

Overconfidence risk.

Standardisation risk.

Prevention, if scaled poorly, can become brittle.

Which is why this section cannot be about expansion alone.

It must be about architecture.

Because what works quietly in one place can fail loudly when copied without care.

Durability, however, introduces a new responsibility.

Once prevention is no longer treated as an exception, it must survive contact with scale. What works quietly in one program, one region, or one leadership cycle is tested when it becomes institutionalised–embedded across portfolios, jurisdictions, and budget horizons. At that point, prevention is no longer protected by novelty. It is exposed to repetition, fatigue, and standardisation.

This is where many organisations stumble again.

They attempt to scale prevention by copying what worked locally, assuming that consistency will preserve effectiveness. Instead, they discover a different risk: brittleness. Early-warning systems become checkbox exercises. Governance processes harden into ritual. Signals lose nuance as they are compressed for reporting. The very mechanisms that once enabled foresight begin to dull it.

The watcher senses this shift before it appears in metrics. Conversations become procedural. Language tightens. Curiosity gives way to compliance. Prevention is still funded–but it is no longer alive.

This is the next threshold institutions must cross.

If prevention is to endure, then it cannot merely be repeated. It must be designed to adapt. It must retain sensitivity without reintroducing panic, authority without reverting to hierarchy, and discipline without erasing judgment. In other words, prevention must be architected–not just governed.

What follows is not an expansion problem, but a systems problem.

How do institutions preserve anticipatory intelligence at scale?

How do they maintain legitimacy without slowing response?

How do they ensure capital remains patient without becoming complacent?

These are not questions of technology or policy alone.

They are questions of structure, ownership, and renewal.

And they mark the point where prevention stops being a program–and becomes an operating model.

Case Study —

The Cost of a Fire That Never Happened: Preventive Power Shutoffs

To prevent wildfires, utilities in California began shutting off power during high-risk conditions.

The logic was simple: no electricity, no ignition.

The impact was immediate. Homes lost power. Businesses paused. Public frustration grew.

And often–nothing happened.

No fire.

No crisis.

Which created the problem.

If nothing happens, then it feels unnecessary.

If something happens, then it feels too late.

Prevention worked–but it looked like overreaction.

There is no visibility for disasters that never occur.

Only cost.

Executive Takeaway:

Prevention succeeds silently–and is judged loudly.

From Pilots to Platforms

Scaling Intelligence When Money Finally Aligns

The moment prevention receives funding does not feel like victory.

It feels tentative.

The watcher notices this immediately–not because the decision was weak, but because of how it was framed. The approval arrived with conditions, milestones, review gates, and sunset clauses. It arrived as permission, not commitment. Prevention has been allowed to exist, but only under supervision. Only as long as it behaves. Only as long as it does not disturb the equilibrium it was designed to protect.

This is not resistance.

It is institutional caution doing what it has always done when confronted with unfamiliar value.

Across organisations, this moment repeats with remarkable consistency. After intelligence gains trust, and after capital finally aligns, prevention is introduced in its smallest acceptable form: a pilot. A contained experiment. A bounded initiative that promises learning without obligation.

Pilots are comfortable.

They signal progress without forcing change. They allow leaders to demonstrate responsiveness while preserving optionality. They generate insight without demanding integration. For new technologies, pilots are often sensible.

For prevention, they are dangerous.

Because prevention that cannot scale is not prevention at all. It is rehearsal.

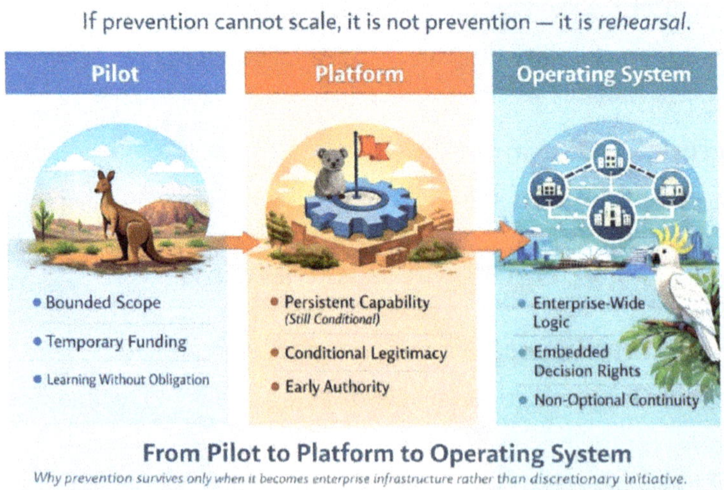

Figure 4.16 Why effective prevention must move beyond pilots and become a non- optional enterprise operating logic

The watcher has seen this before. Capable systems, proven under controlled conditions, gradually fade once the pilot phase ends. Not because they failed, but because no one designed the organisation to carry them forward. Ownership remained unclear. Budgets remained provisional. Governance treated the capability as an exception rather than infrastructure.

This is where many prevention efforts quietly die–not in opposition, but in ambiguity.

The Pilot Trap

Pilots succeed by isolating complexity.

Platforms survive by absorbing it.

The distinction matters.

A pilot proves that something can work. A platform ensures that it does–repeatedly, reliably, and without constant justification. Moving from one to the other requires a shift that is less technical than institutional.

The watcher observes the early signs of strain as the pilot begins to mature. The system performs as expected. Signals arrive early. Decisions are better informed. Volatility softens at the edges. And yet, questions begin to surface.

Who owns this once the pilot ends?

Which budget absorbs it next year?

How does this integrate with existing operational cadence?

What happens when leadership changes?

These are not implementation questions. They are survival questions.

Prevention systems fail when they remain framed as projects. Projects have timelines. They end. Prevention does not. Risk does not respect project boundaries or funding cycles. A system designed to watch continuously cannot be governed episodically.

This is the first architectural challenge of scale: continuity without novelty.

Pilots benefit from attention. Platforms must endure without it.

As the pilot runs quietly–without incidents, without escalation, without headlines–the organisation begins to experience an uncomfortable contradiction. The system is working, but its success produces no visible justification for expansion.

The watcher hears this expressed indirectly:

"Do we need this everywhere?"

"Are the benefits strong enough to justify enterprise rollout?"

"Could we concentrate this capability only in the highest-risk areas?"

These are reasonable questions. They are also evidence that the organisation is still evaluating prevention through the lens of exception rather than baseline.

What the watcher understands–and what the institution is still learning–is that prevention does not demonstrate value through intensity. It demonstrates value through absence of variance. That absence becomes harder to defend as it becomes normal.

The more stable the system feels, the easier it is to argue that stability would have existed anyway.

This is the paradox of successful anticipation: it erases the contrast that made it persuasive.

Scaling requires a different kind of evidence–not proof that disaster was imminent, but proof that governance has changed posture.

The critical moment does not arrive as a rejection. It arrives as delay.

A proposal to extend the system beyond the pilot footprint circulates. It is stronger than the first. The economics are clearer. The governance model is more mature. Insurance signals are supportive. Operational teams advocate quietly but consistently.

The decision stalls.

Not because of disagreement, but because of sequencing.

"We should consolidate learnings first."

"Let's ensure we don't over-engineer."

"We can revisit this next cycle."

The watcher recognises the pattern instantly. This is not indecision.

It is the organisation waiting for permission to make the capability permanent.

Permanence is costly.

It requires admitting that prevention is no longer optional. That early intelligence is no longer advisory. That some decisions must be designed to occur before justification is obvious.

Institutions hesitate at this threshold because permanence redistributes responsibility.

Once a system becomes a platform, failure is no longer attributable to experimentation. It becomes attributable to design.

From Capability to Operating Model

This is the inflection point Section 4.3 addresses.

The problem is no longer whether prevention works, or even whether it can be funded. The problem is whether the institution can redesign itself to carry foresight without reverting to reaction.

Scaling prevention demands answers to questions most organisations avoid:

Where does anticipatory intelligence live structurally?

How is early action authorised without escalation theatre?

How is success reviewed when nothing happens?

How is capital protected from being clawed back during quiet periods?

How does leadership change without resetting legitimacy?

These are not AI questions. They are operating-model questions.

The watcher sees that the organisation is standing at a familiar boundary. On one side is a well-run pilot–effective, admired, but fragile. On the other is a platform that would quietly reshape decision-making across functions, budgets, and time horizons.

Crossing that boundary requires more than scale. It requires architecture.

This section is not about expansion for its own sake. It is about designing prevention so that it survives success, leadership change, and institutional memory loss.

It is about moving from:

- isolated intelligence → embedded perception
- pilot funding → structural capital
- heroic justification → routine stewardship

And it is about recognising a final truth the watcher now understands clearly: Prevention does not fail when it is rejected. It fails when it is adopted without being redesigned to endure.

What follows is the anatomy of that redesign.

Not as theory. As practice.

Designing for Permanence

The moment an organisation decides to move beyond pilots, it encounters a problem it rarely names directly: prevention does not fit comfortably anywhere.

It is not purely operational, because it does not act on a fixed schedule.

It is not purely strategic, because its outputs are continuous rather than episodic.

It is not purely technological, because its value depends less on precision than on trust. When prevention is placed inside any single function, it inherits that function's limits.

The watcher sees this play out as the pilot transitions into its second year. The question of "where this lives" becomes unavoidable. Operations wants influence, because they feel the friction first. Risk wants oversight, because escalation logic runs through them. Technology wants stewardship, because the system is complex and evolving. Finance wants visibility, because capital has now been committed.

Each claim is reasonable. None is sufficient on its own.

This is the first architectural lesson of scale: anticipatory systems cannot be owned the way reactive systems are. They do not belong to the part of the organisation that acts last. They belong to the part that decides when acting early is allowed.

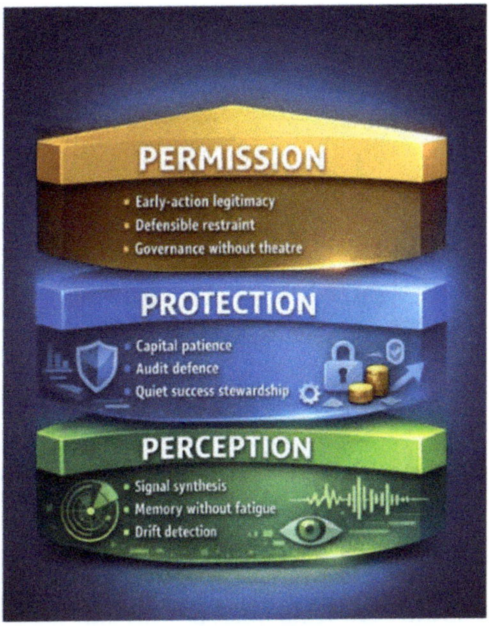

Figure 4.17 Prevention endures not through central control, but through the separation of perception, protection, and permission

The Platform Principle

What survives is not a department, but a platform.

A prevention platform is not a single system. It is a set of persistent capabilities that operate together, regardless of who is in the room or what incident is dominating attention. At its core, the platform does three things continuously:

- It observes patterns across domains without forgetting.
- It translates those patterns into decision-relevant signals.
- It holds the authority to legitimise early action without escalation theatre.

These capabilities must remain intact even as tools, vendors, and leadership change.

The watcher notices that the organisations that scale successfully stop asking, "Who runs this system?" and begin asking, "Who is accountable for its continuity?"

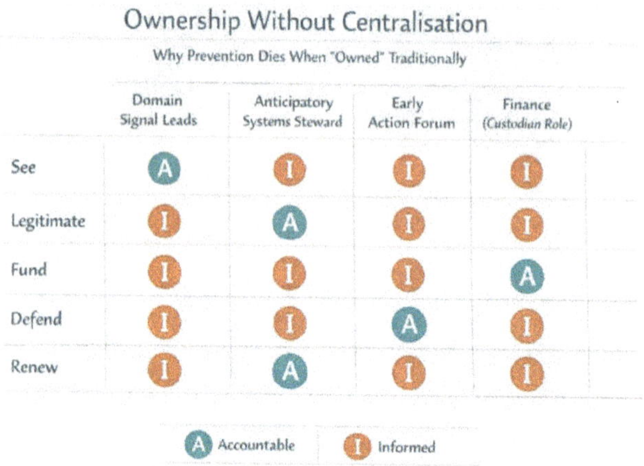

Figure 4.18 Prevention survives when authority is distributed by function, not centralised by ownership

That shift matters.

Systems can be run. Platforms must be carried.

Separation of Powers, Not Centralisation

One of the most common scaling failures occurs when organisations attempt to centralise prevention too aggressively. In an effort to control quality and consistency, they pull all anticipatory intelligence into a single unit.

The result is predictable.

Signals lose context.

Local nuance is flattened.

Response teams disengage.

Prevention becomes technically impressive and operationally irrelevant.

The alternative is not decentralisation, but separation of powers.

In organisations that get this right, ownership is deliberately split:

- Signal stewardship sits close to the data and domain expertise
- Decision legitimacy sits within governance and risk authority
- Capital protection sits with finance, not as approver but as custodian

No single group controls the entire chain.

This architecture prevents capture. It ensures that early signals cannot be suppressed locally, nor exaggerated centrally. It also distributes responsibility for early action so that no one individual must carry the reputational risk alone.

The watcher realises that this is what the organisation has been missing all along: not courage, but structure.

The Role That Did Not Exist Before

As the platform takes shape, a new kind of role emerges–not always formally titled, but functionally distinct.

This role is not a data scientist.

It is not a risk officer. It is not a program manager.

It is the steward of anticipatory legitimacy.

This person–or small group–is accountable for ensuring that early signals remain interpretable, defensible, and actionable over time. They do not decide outcomes. They maintain the conditions under which early decisions can be made without apology.

They track drift.

They monitor threshold fatigue.

They ensure that silence is understood, not ignored.

When leadership changes, this role provides continuity. When budgets tighten, it protects the platform from being quietly dismantled. When incidents occur, it ensures that prevention is not blamed for what it was never designed to stop.

The watcher sees how transformative this role is–not because it adds power, but because it absorbs pressure.

Funding Without Fragility

Scaling also forces a reckoning with capital design.

Pilot funding is always temporary. Platform funding cannot be.

The organisations that succeed stop treating prevention as discretionary spend and begin treating it as baseline infrastructure, comparable to safety systems, compliance mechanisms, or core analytics. This does not mean funding becomes untouchable. It means cuts must be explicit and directional.

Quiet erosion is no longer allowed.

Finance plays a different role here. Instead of asking whether prevention "paid for itself," it monitors indicators of degradation:

Are response costs creeping back?

Is insurance patience tightening?

Are volatility bands widening?

Is decision latency increasing?

These signals replace incident counts as the health metrics of the platform.

The watcher recognises this as the moment capital finally becomes aligned with foresight–not because risk has increased, but because confidence in governance has matured.

Cadence Over Crisis

Finally, scale demands rhythm.

Reactive systems move to the cadence of incidents. Anticipatory platforms require a cadence independent of crisis.

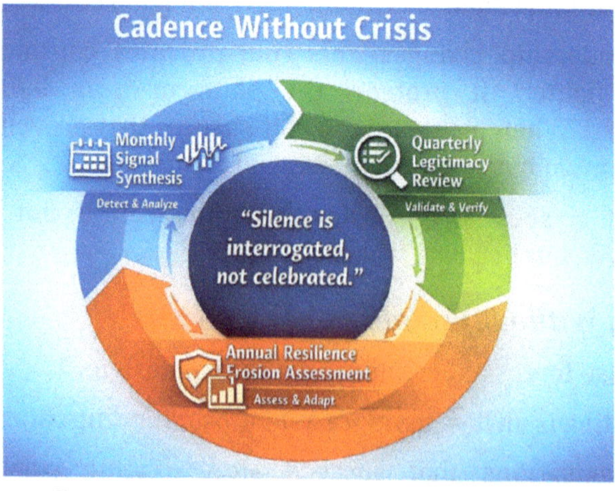

Figure 4.19 ***Effective prevention is not passive. It is governed through disciplined, recurring scrutiny–even when nothing happens.***

Regular reviews occur whether or not anything has happened. Patterns are examined even when thresholds remain unbroken. Silence is interrogated, not celebrated.

This is uncomfortable at first.

Meetings without urgency feel indulgent. Discussions without decisions feel inefficient. But over time, the organisation learns that cadence is what prevents panic.

The watcher sees this shift most clearly when a new anomaly appears and does not trigger chaos. The platform absorbs it. The signal is contextualised. Options are preserved. Nothing dramatic occurs.

And that, quietly, is the point.

Where this Leaves the Organisation

By the time prevention is fully platformed, the organisation has changed in ways that are difficult to summarise in a slide.

It reacts less. It escalates later. It decides earlier.

Most importantly, it no longer requires fear to justify foresight.

This is not because risk has diminished. It is because the institution has learned how to live with it–structurally, economically, and without spectacle.

What remains is the final test.

Whether this architecture can survive beyond the individuals who built it.

That question does not belong to Section 4.3 alone.

It belongs to the operating model that follows.

Designing an Operating Model that can Carry Foresight

By the time the organisation accepts that prevention must scale, the most dangerous illusion has already been dispelled.

Technology was never the hard part.

The watcher understands this now with a clarity that would have felt impossible earlier in the story. Models improved. Signals converged. Capital hesitated, then arrived. Governance shifted from scepticism to conditional permission.

And still–everything feels fragile.

Because the institution has not yet answered the question that determines whether prevention survives.

This is the point where many organisations fail quietly. They fund prevention without redesigning themselves to carry it. They scale tools without scaling authority. They add intelligence without altering decision rights.

The result is not rejection. It is slow reversion.

This Section exists to prevent that.

The Operating Model Problem

Institutions are excellent at absorbing new capabilities into old structures. It is how they remain stable.

But stability is precisely what makes anticipatory systems vulnerable. When prevention is absorbed without structural accommodation, it is evaluated by criteria it was never designed to meet.

- *Why didn't it stop this incident?*
- *Why didn't it escalate sooner?*
- *Why are we still funding this if nothing happened?*

These questions are not malicious. They are inevitable in organisations optimised for reaction.

The watcher has seen this cycle play out repeatedly: a system works quietly, legitimacy erodes slowly, and eventually prevention is reframed as "nice to have." Not because it failed–but because no one protected its posture.

What follows is the operating model that prevents that erosion.

Principle One: Foresight Is Not a Function

The first mistake organisations make is trying to "place" prevention.

They ask:

Is this a risk function?

A technology function?

An operations function?

A strategy function?

The answer is no.

Prevention is not a function. It is a cross-cutting capability that modifies how all functions behave under uncertainty.

When it is owned like a function, it inherits that function's incentives:

- operations optimise throughput
- risk optimises defensibility
- technology optimises performance
- finance optimises cost

None optimise timing.

Timing is the core value of prevention–and timing belongs nowhere unless it is explicitly designed.

Principle Two: Separate Perception, Permission, and Protection

The organisations that scale successfully do something subtle and powerful: they separate the responsibilities that were previously collapsed into escalation.

Instead of asking one group to see, decide, and defend early action, they distribute those burdens deliberately.

1. **Perception Layer — Seeing Without Acting**

   This layer is responsible for continuous signal synthesis.

   It:

   - integrates environmental, infrastructure, operational, and external data
   - tracks persistence, interaction, and drift
   - remembers what humans normalise

   It does not decide action.

   This insulation is critical. The moment perception becomes accountable for action, it becomes conservative. Signals soften. Language hedges. Early warnings become recommendations rather than observations.

The watcher understands this instinctively. They have lived inside it.

Perception must be allowed to speak without consequence.

2. **Permission Layer — Acting Without Drama**

This layer exists to do something institutions are rarely designed to do: authorise early action without requiring crisis language.

It owns:

- decision thresholds that are temporal, not just quantitative
- legitimacy rules for acting under uncertainty
- escalation without spectacle

This is where governance lives–not as gatekeeper, but as translator.

The permission layer does not ask: "Is this proven?"

It asks: "Is this defensible if nothing happens?"

That distinction changes everything.

3. **Protection Layer — Defending Quiet Success**

The final layer protects prevention from its greatest enemy: hindsight.

This layer:

- tracks avoided volatility
- monitors resilience erosion
- defends funding during quiet periods
- absorbs audit and scrutiny pressure

Finance plays a central role here–not as approver, but as custodian of optionality.

The watcher sees the shift when finance stops asking for ROI and starts asking for early warning of governance degradation.

That is maturity.

Roles That Make This Work

Operating models fail when roles are ambiguous. Prevention fails when roles are absent.

1. The Anticipatory Systems Steward

This role did not exist before–and most organisations still do not name it. It is not a technologist.

Not a risk officer.

Not a strategist.

It is the keeper of continuity.

They are accountable for:

- ensuring perception remains independent
- ensuring permission remains legitimate
- ensuring protection remains active

They do not decide outcomes.

They prevent collapse into reaction.

The watcher recognises this role immediately–because it formalises what they have been doing informally for years.

These roles sit close to reality.

They:

- understand local context
- maintain signal fidelity
- resist pressure to normalise drift

They do not escalate.

They inform.

Their authority comes from memory, not hierarchy.

This is not an emergency committee.

It meets regardless of incidents.

Its purpose is not to decide what happened, but to decide whether early intervention is now legitimate.

Membership is stable.

Language is disciplined.

Decisions are reversible but documented.

The watcher notices how much power this forum absorbs–simply by existing.

2. Decision Rights That Matter

Most organisations rely on escalation as a proxy for authority.

Prevention cannot survive escalation-only models.

Instead, decision rights are redesigned around windows, not events.

- *Who can authorise limited intervention before thresholds breach?*
- *Who can allocate contingency capital without declaring emergency?*
- *Who can delay visible response to preserve optionality?*

These rights are explicit.

Silence is not consent.

Waiting is not neutral.

This clarity removes the personal risk that previously paralysed early actors. Reactive systems move to incident tempo.

Preventive platforms require institutional rhythm.

- Monthly signal synthesis
- Quarterly legitimacy review
- Annual resilience erosion assessment

Nothing dramatic happens in these forums.

That is precisely why they work.

The watcher sees how cadence replaces vigilance with discipline.

3. Funding as Architecture

Platform funding is not project funding.

It is:

- Multi-year
- Baseline
- Explicitly protected

Cuts, when they occur, must be justified in terms of lost capacity, not saved cost.

Finance begins tracking:

- decision latency
- volatility bands
- forced vs chosen actions

These become leading indicators of governance health.

4. Audit Without Punishment

Audit often kills prevention unintentionally.

In this model, audit is re-positioned.

Auditors ask:

- *Were early decisions documented?*
- *Was silence interrogated?*
- *Were thresholds treated as guidance, not absolutes?*

Audit validates discipline–not outcomes.

The watcher realises how powerful this is. It removes fear from foresight.

5. Surviving Leadership Change

This is the final test.

When leadership turns over, most prevention systems reset.

In this model:

- roles persist
- forums persist
- cadence persists

Prevention survives not because leaders believe in it–but because it has become invisible infrastructure.

6. The Quiet Outcome

By the time this operating model is fully embedded, the organisation feels different.

Not safer. Not complacent.

7. **Calmer under Uncertainty**

Decisions take longer when they should–and faster when they must. Escalation becomes rarer, but more decisive. Capital moves earlier, but with less drama.

The watcher understands the irony: The better prevention works, the less it is noticed.

That is no longer a vulnerability.

It is the proof that the system has learned how to carry foresight without requiring catastrophe to legitimise it.

Prevention does not scale through conviction. It scales through architecture.

Not belief–but design.

Not heroics–but roles.

Not urgency–but rhythm.

This is how intelligence becomes governable.

This is how capital learns to wait.

This is how institutions stop mistaking reaction for responsibility.

The Point Where Architecture Stops Being Optional

The organisation does not recognise the moment when prevention becomes fragile again.

That is the danger of success.

The pilot scaled. The platform held. Early action became defensible. Capital learned to wait. Governance adapted just enough to carry foresight without requiring drama. From the inside, this feels like progress–careful, earned, and mature.

From the watcher's vantage point, it feels incomplete.

Not wrong. Not broken. But unfinished.

Because everything that has been built so far works precisely because it is still contained.

Contained within a region.

Contained within a function.

Contained within a leadership cohort that understands how it came to be.

The watcher sees what others do not yet name: prevention has learned how to survive locally, but it has not yet learned how to survive at scale.

And scale changes everything.

When Local Success Creates Systemic Risk

The first sign is subtle.

Different parts of the organisation begin implementing similar capabilities–but not identically. One region adapts escalation rules. Another modifies thresholds. A third adjusts cadence to suit operational rhythm. None of these choices are unreasonable. Each reflects local context.

Collectively, they introduce drift.

Signal interpretation begins to vary.

Permission logic fragments.

Capital protection becomes uneven.

Prevention still works–but not the same way everywhere.

The watcher recognises this pattern immediately. It is the same one that appears whenever institutions scale without coherence. What once felt like resilience becomes variance. What once felt like flexibility becomes ambiguity.

No one has done anything wrong.

That is the problem.

Leadership responds predictably.

"Let's replicate what worked."

"Let's codify best practice."

"Let's standardise templates."

Replication feels safe. It suggests control.

But prevention is not a product to be copied. It is a posture to be carried.

When replication happens without architecture, it produces imitation without legitimacy. Teams follow the motions without inheriting the reasoning. Decisions are made earlier–but not always more wisely. Escalation happens faster–but not always more defensibly.

The watcher notices something unsettling: the organisation has begun producing early action again–but this time without the careful legitimacy that made it possible in the first place.

The pendulum has started to swing.

What worked locally relied on a set of unspoken agreements: That finance would defend quiet success.

That governance would absorb scrutiny.

That operations would trust early signals.

That leadership would resist the urge to intervene theatrically.

These agreements were never written down.

They were lived.

At scale, lived agreements collapse.

New leaders arrive without memory. New regions inherit capability without context. New audits apply old logic to new behaviour. Prevention becomes visible again–and visibility attracts challenge.

The watcher sees the questions returning in familiar forms:

"Who authorised this?"

"Why wasn't this escalated further?"

"What precedent does this set?"

The organisation is not backsliding. It is encountering the limits of local architecture.

The Moment Prevention Becomes an Enterprise Problem

This is the quiet inflection point where prevention stops being a success story and becomes an enterprise design problem.

Not because it has failed. But because it has succeeded enough to matter.

The watcher realises something profound in this moment: the organisation has learned how to act early–but it has not yet learned how to agree early.

Agreement at scale requires more than tools, funding, or even governance forums. It requires a shared logic that travels intact across functions, regions, leadership cycles, and scrutiny environments.

Without that logic, prevention fragments.

Fragmentation does not announce itself as failure. It announces itself as inconsistency. And inconsistency is what institutions fear most–because it undermines defensibility.

Why Centralisation Will Not Save This

The instinctive response is to centralise.

Create an enterprise prevention office.

Mandate a single escalation framework.

Enforce uniform thresholds.

The watcher has seen this approach before.

It works briefly. Then it collapses under its own weight.

Centralisation strips context from perception. It replaces local nuance with generic signal. It forces early action to compete with enterprise politics. It recreates the very bottlenecks prevention was meant to dissolve.

Worse, it concentrates reputational risk.

When prevention is centralised, every early decision becomes a visible act of authority. Silence becomes suspicious. Action becomes political.

The organisation senses this risk intuitively and hesitates.

Why Decentralisation Will Also Fail

The alternative is to trust autonomy.

Let regions adapt.

Let functions decide.

Let local leaders own early action.

This preserves context–but sacrifices coherence.

Over time, the same signal produces different responses in different places. What is legitimate in one region is questioned in another. Capital behaves unevenly. Audits escalate selectively.

Prevention begins to look arbitrary.

The watcher understands the bind now clearly: scale demands both coherence and context–and institutions are rarely designed to hold both at once.

The breakthrough does not arrive through analysis.

It arrives through discomfort.

A senior leader–thoughtful, pragmatic–voices what others have been circling: "We're not struggling with intelligence. We're struggling with alignment."

The room quiets.

Alignment, not accuracy.

Coherence, not capability.

Continuity, not conviction.

This reframes everything.

Because alignment is not a technology problem. It is an enterprise architecture problem.

What the Organisation Finally Sees

At this point in the journey, the organisation can name what it has been missing all along.

Not another system.

Not another model.

Not another governance layer.

What it lacks is a shared continuum–a way to connect:

- How risk is perceived
- How legitimacy is granted
- How capital is protected

- How action is authorised
- How outcomes are reviewed

...without resetting at every boundary.

The watcher understands that prevention has reached the limits of episodic design. It can no longer survive as a collection of policies, pilots, or platforms.

It needs a unifying logic.

Why Frameworks Usually Fail Here

Most frameworks arrive too early or too late.

Too early, and they feel theoretical.

Too late, and they feel imposed.

What is required now is not instruction–but articulation.

The organisation does not need to be told what to do. It needs a way to describe what it has already learned–in a form that can be carried forward intact.

This is the difference between a framework and a continuum. Frameworks prescribe.

Continuums connect.

The Shape of What Comes Next

The watcher can sense the contours of what is forming, even if it has not yet been named.

A structure where perception, permission, and protection are not functions–but phases.

A model where early intelligence does not beg for legitimacy–it inherits it.

A system where capital does not arrive late–it is designed to wait.

Most importantly, a logic where prevention does not have to argue its worth repeatedly–because its absence would be immediately felt.

This is no longer about wildfire, infrastructure, or any single hazard. It is about how institutions govern uncertainty itself.

Why This Cannot Remain Implicit?

As long as this logic remains implicit, it will erode.

New leaders will reinterpret it.

New crises will override it.

New pressures will fragment it.

Implicit wisdom does not scale.

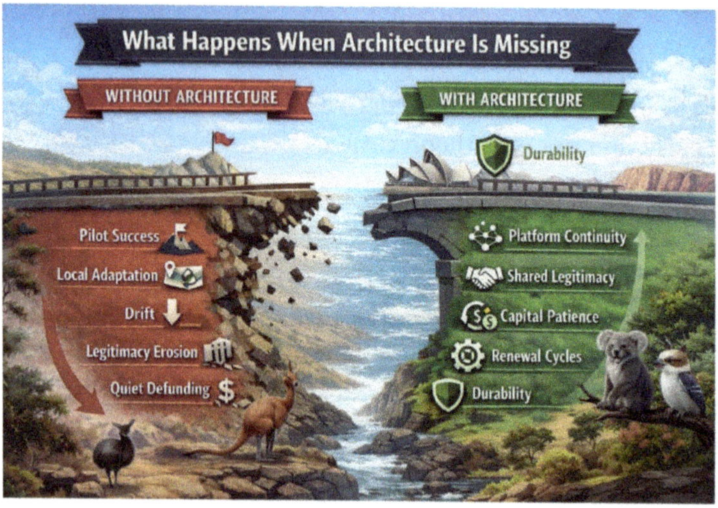

Figure 4.20 Without architecture, success erodes quietly. With architecture, success endures

The organisation understands now, that if it wants foresight to survive beyond individuals, it must be made explicit–not as policy, but as architecture.

This is where the narrative must cross a line.

Up to now, the story has been about learning to see, trust, fund, and carry prevention.

From here on, the story must be about designing institutions that do not need to relearn those lessons every cycle.

The watcher understands the significance of this moment.

For the first time, the organisation is not asking: "How do we improve prevention?"

It is asking: "How do we redesign ourselves so prevention does not have to fight for survival again?"

That question changes everything.

Prevention has learned how to survive locally.

It has learned how to justify itself economically.

It has learned how to earn trust operationally.

What it has not yet learned is how to exist everywhere at once without losing coherence.

That is no longer a question of technology, finance, or governance alone.

It is a question of enterprise design.

And that is where the next chapter begins.

What This Chapter Established

Why Prevention Finally Becomes Durable

Prevention failed under traditional ROI logic because it produced quiet success. Reframed as volatility reduction, optionality, and resilience capital, prevention became an asset rather than an argument–something to steward, not defend.

Executive Takeaways — What This Means for Leaders

- Prevention fails under traditional ROI because success leaves no visible trace.
- Value must be reframed in terms of resilience, volatility reduction, and optionality.
- Capital does not follow prediction–it follows clarity of decision logic.
- Early action is economically viable when institutions recognise the cost of waiting.
- Prevention must be funded as a system capability, not justified as an exception.

What Comes Next

The Responsibility of Acting Early

Once foresight is governable and fundable, the final question is no longer whether institutions can act early–but whether they are willing to carry that responsibility at scale.

Chapter 5
The Responsibility of Foresight

Leadership in an Age of Permanent Risk

By this point, the argument no longer rests on technology.

The preceding chapters have established something more fundamental: that the failure of prevention is not caused by weak intelligence, insufficient data, or immature tools. It is caused by institutional structures that grant authority too late, reward certainty over restraint, and fund response more easily than foresight.

Artificial Intelligence has already altered the timeline of risk. Signals now arrive earlier than institutions are prepared to legitimise. Patterns persist long before thresholds are crossed. Convergence emerges while governance still waits for events.

This creates a new condition for leadership.

Responsibility no longer begins when damage occurs.

It begins when signals persist and nothing has happened yet.

This chapter closes the book by addressing what that shift demands from leaders, boards, and institutions. It does not introduce new frameworks or technologies. It translates what has already been established into institutional posture.

The responsibility of foresight is not prediction. It is governance before consequence.

Why Ecosystems Are Intelligence Systems

Wildlife and land as early-warning infrastructure

For most of modern organisational history, intelligence has been defined narrowly: information generated inside institutional boundaries, formatted for governance consumption, and validated through familiar reporting artefacts.

The Animal That Moved First

In the days before conditions worsened,
wildlife had already begun to shift.

A field observer noticed it first.

Animals were moving away–not suddenly, not in panic, but gradually, deliberately.

There were no alerts for this.
No dashboards capturing it in real time.

But the signal was there.

Embedded in behaviour, not data.

By the time systems registered the risk, the landscape had already begun to change.

Nature had responded earlier than the institution.

Figure 5.1 Nature speaks early. Intelligence decides whether to listen

Ecosystems did not qualify.

Land, wildlife, and environmental systems were treated as external conditions–factors to be managed after disruption, insured against, or restored once damaged. Their signals were considered background context, not decision-grade intelligence.

That distinction is no longer viable.

Long before a fire ignites, ecosystems register stress. Vegetation dries unevenly. Soil moisture deviates from seasonal norms. Wildlife alters movement patterns. Wind behaviour becomes erratic. Infrastructure corridors experience minor, self-clearing faults.

None of these constitute an event.

All of them describe direction.

What has changed is not the behaviour of ecosystems, but the capacity of institutions to listen to them. Artificial Intelligence, at its most useful, does not predict catastrophe. It preserves memory. It holds continuity across time. It notices persistence where humans normalise repetition and institutions dismiss unease.

In doing so, AI converts ecological response into legible intelligence.

Wildlife movement is no longer only a conservation concern; it becomes an indicator of habitat stress, thermal pressure, and future ignition risk. Land degradation is no longer only an environmental issue; it becomes a precursor to infrastructure exposure, insurance repricing, and operational fragility.

When institutions treat ecosystems as passive assets, they forfeit the earliest layer of intelligence available to them. When they treat ecosystems as sensing systems, they gain time–and with time, optionality.

This is not a philosophical shift. It is a governance one.

The responsibility of foresight begins with acknowledging that the earliest warnings about disaster rarely originate inside boardrooms, dashboards, or control centres. They emerge from the systems institutions depend upon–whether governance is prepared to recognise them or not.

Case Study —

When Nature Solved What Systems Couldn't: Yellowstone Wolves

When wolves were reintroduced into Yellowstone, it wasn't a technological intervention. It was a simple ecological decision.

But the impact was profound.

Deer behaviour changed. Vegetation recovered. Rivers stabilised. The entire ecosystem began to rebalance itself.

No central system designed this outcome. Nature adjusted on its own.

Ecosystems don't wait for thresholds. They respond continuously.

The lesson is simple: intelligence doesn't always come from control–sometimes it comes from restoring balance.

Executive Takeaway:

Resilience is not always engineered–it is often restored.

Institutions That Learn Will Last

Adaptive Governance as a Survival Strategy

Learning is often invoked as a virtue of resilient organisations. In practice, however, most institutional learning remains retrospective. Reviews follow incidents. Lessons are extracted from failure. Controls are adjusted after damage has already occurred.

This is not adaptive learning. It is post-event correction.

In environments where risk accumulates quietly and converges across systems, learning that occurs only after consequence is structurally late.

Adaptive institutions learn before failure.

They do not wait for loss to validate intelligence. They adjust decision rights when signals persist. They recalibrate governance when patterns emerge, even if outcomes remain quiet.

The earlier chapters showed why this is difficult. Institutions were designed for episodic risk–environments where thresholds were reliable, signals were noisy, and acting too early carried disproportionate reputational and financial cost.

That design logic no longer holds.

Today, institutions that rely on event-driven learning are not cautious. They are misaligned with the shape of modern risk.

Adaptive governance requires three conditions.

First, learning must be continuous. Intelligence does not arrive as a single moment of clarity. It accumulates through persistence, recurrence, and interaction. Institutions that treat learning as an episodic exercise miss the period when intervention remains reversible and affordable.

Second, learning must be legitimate. Information that cannot be acted upon safely is not intelligence; it is noise. Adaptive institutions design permission pathways that allow early signals to influence decisions without demanding certainty or crisis justification.

Third, learning must be protected. When early action succeeds, nothing happens. Without explicit protection from budget erosion and governance amnesia, prevention collapses back into pilots and proofs.

Institutions that fail to meet these conditions rarely collapse suddenly. They endure–until margins vanish and options narrow.

Endurance is often mistaken for resilience.

In reality, it is frequently just deferred exposure.

The responsibility of foresight requires leaders to move beyond learning as reflection and toward learning as structural adaptation–not as culture, but as governance.

Institutions that learn early will last.

Institutions that learn late will explain.

Chapters 1 through 5 establish the practitioner case for governing foresight as an institutional responsibility–spanning intelligence, legitimacy, capital, and leadership.

Listening as Leadership

What It Now Means to Lead Responsibly

For decades, leadership competence has been measured by decisiveness under pressure. The ability to act visibly during crisis became synonymous with authority, courage, and control.

That model is no longer sufficient.

In an environment where intelligence arrives early and consequence arrives late, leadership is no longer defined by reaction. It is defined by listening.

Listening, in this context, is not consultation or consensus-building. It is sustained attention to signals that do not yet justify action in conventional terms. It is the willingness to hold ambiguity without dismissing it simply because it lacks a headline.

This form of leadership is uncomfortable.

Listening exposes leaders to uncertainty. It demands decisions without guarantees. It requires standing behind restraint when nothing happens–and defending that restraint to boards, auditors, regulators, and markets accustomed to rewarding visible action.

The temptation, in such moments, is to delay. To request further validation. To wait for a clearer signal.

But waiting is no longer neutral.

Leadership today is measured not by how well crises are managed, but by whether they are quietly avoided without eroding legitimacy.

Responsible leadership now includes:

- Authorising early action without event validation,
- Protecting teams who act on persistent signals,
- And sustaining investment in capabilities whose success produces no visible outcome.

It also includes knowing when not to act–and being able to explain that decision later without embarrassment or defensiveness.

Listening as leadership is not passivity. It is disciplined restraint, supported by architecture, governance, and capital design.

Leaders who cannot listen early will eventually be forced to act late.

Leaders who can listen early must be prepared to explain why nothing happened.

That explanation–calm, defensible, institutional–is now a core leadership skill.

Figure 5.2 Authority now comes from what leaders choose to hear early, not what they decide under pressure

Designing What Comes Next

From Foresight to Durable Action

The purpose of foresight is not prediction. It is durability.

Durable prevention does not emerge from better models alone. It emerges when intelligence, legitimacy, and capital are designed to reinforce one another across normal operations, budget cycles, and leadership transitions.

The earlier chapters demonstrated what happens when this architecture is missing. Pilots succeed, then erode. Platforms gain attention, then become optional. Prevention works, then quietly disappears.

Designing what comes next requires accepting a final, uncomfortable truth: Prevention cannot be justified continuously. It must be governed continuously.

This is where foresight becomes responsibility.

Figure 5.3 Durability is not achieved by reacting faster, but by acting early–and making restraint sustainable

Importantly, this transition from insight to durability is not merely conceptual. It is already being attempted in practice–not through generic platforms or research prototypes, but by embedding anticipatory intelligence into architectures that governance can recognise and capital can sustain.

The system designs reflected in the patented approaches underlying this work were not conceived as prediction engines or technical novelties. They were structured deliberately as institutional instruments–mechanisms intended to translate early signals from land, infrastructure, and ecological systems into decision pathways that can be authorised, funded, and defended over time.

What matters here is not the existence of patents themselves, but what their structure reveals: a recognition that intelligence only creates impact when it is packaged in forms institutions are prepared to own. By anchoring anticipatory AI within explicit governance logic and economic rationale, these designs attempt to address the core failure explored throughout this book–not detection, but durability.

In that sense, they serve as practical demonstrations of the book's central claim: that profitable fire prevention and wildlife protection emerge not from better prediction alone, but from architectures that allow early restraint to survive normal operations, scrutiny, and institutional memory loss.

In fire prevention and wildlife protection, foresight is no longer optional expertise–it is leadership duty. When AI reveals emerging ecological risk early, institutions are no longer deciding whether to act, but whether they accept responsibility for what follows if they do not. The future will judge governance not by response speed, but by whether it acted while prevention was still possible.

What comes next, therefore, is not another framework or platform. It is a choice.

A choice to treat early intelligence as an asset, not an input.

A choice to treat restraint as performance, not indecision.

A choice to interrogate silence rather than celebrate it.

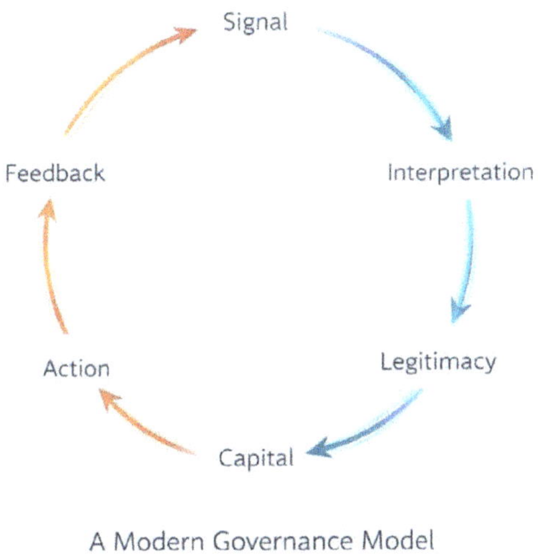

Figure 5.4 From signal to action–governance is a continuous loop, not a linear step

Executive Takeaways — What This Means for Leaders

- Leadership now begins before outcomes–not after them.
- Ecosystems are not external conditions. They are intelligence systems.
- Institutions that learn early will outlast those that correct late.
- The responsibility to act emerges when signals persist–not when events force agreement.
- Foresight is not optional. It is the new baseline of accountability.

The Leader Who Acted Without an Event

A senior leader reviewed the signals carefully.

There was no incident.
No escalation. No external pressure.

Only a pattern–subtle, persistent, and increasing.

The decision to act came early.

Operations were adjusted. Exposure reduced.

No announcement was made.
No urgency declared.

Nothing happened.

No recognition followed.

But months later, when similar conditions led to failure elsewhere, the difference became clear.

Leadership, in that moment, was not defined by response.

It was defined by the willingness to act before justification arrived.

Case Study —

Designing Before Disaster: The Netherlands and Living with Water

In the Netherlands, flooding is not a possibility–it's a constant risk.

Instead of only building higher barriers, the country changed its approach. It created space for water through the "Room for the River" program.

Land was redesigned. Cities adapted. Systems were built assuming uncertainty–not stability.

The key shift was this:

They stopped waiting for disaster to justify action.

They designed for what could happen–before it did.

In some systems, waiting for proof is the real risk.

Executive Takeaway:

The most resilient systems act before certainty arrives.

Closing Reflection — The Final Responsibility

Disasters are no longer sudden.

Risk no longer announces itself politely.

And leadership can no longer wait for permission from consequence.

The responsibility of foresight is not to eliminate uncertainty or predict perfectly. It is to design institutions that can act early, defend restraint, and sustain prevention–even when success looks like nothing happened.

That responsibility now rests with leaders.

Not because technology demands it, but because reality does.

Those who accept it will rarely be recognised for what did not occur.

They will, however, leave behind institutions that last.

And in an age of permanent risk, that may be the highest form of leadership available.

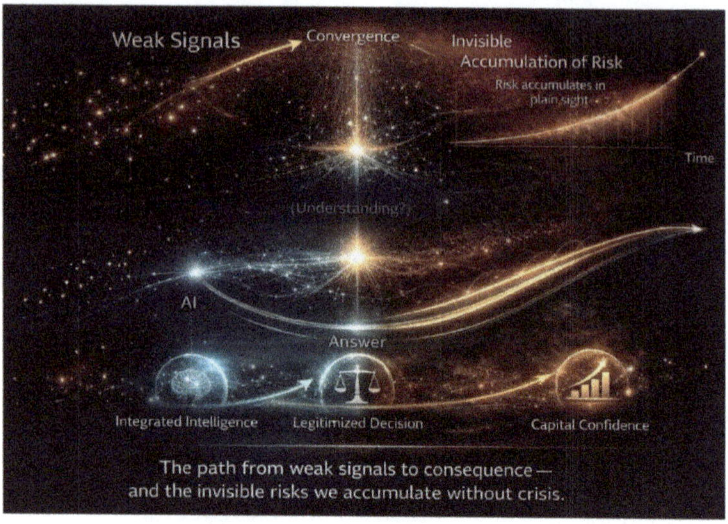

Figure 5.5 Risk does not explode–it accumulates, unseen, until it cannot be ignored

What This Chapter Established

When Early Knowledge Becomes Obligation

In a world where risk is visible long before catastrophe, inaction is no longer neutral. Foresight becomes a leadership responsibility–toward institutions, ecosystems, and the future itself.

What This Book Leaves You With

Capacity, Not Certainty

The future will not be decided in the moment of crisis. It will be decided in the quiet moment before it–when the signal was visible, and the choice to act was still optional.

That moment rarely feels important. But it is.

Executive Synthesis — The New Logic of Prevention

- Signals appear early. Legitimacy arrives late. This gap defines modern risk.
- Intelligence alone does not create impact. Governance determines whether it is used.
- Prevention is not a technical problem. It is an institutional design challenge.
- The cost of waiting is real–even when it does not appear on balance sheets.
- The future will not be shaped by what we can predict–but by what we are willing to act on before it demands proof.

Afterword

By Dr. Dakshinamurthy Venkata Kolluru

The Courage to Act Before the Fire

Every book eventually reaches a quiet turning point. The evidence has been laid out. The frameworks are explained. The stories have been told. What remains is a simpler and more difficult question:

What Will We Actually Do With What We Now Understand?

Throughout this book, we have explored a paradox that increasingly defines modern institutions. Our world has become more visible than ever before. Sensors monitor landscapes continuously. Satellites observe environmental change at planetary scale. Data flows across infrastructures, markets, and ecosystems in near real time. Artificial Intelligence now detects subtle patterns and converging signals long before human systems traditionally recognize them.

In many ways, we are living in the most informed era in history.

And yet, despite this unprecedented visibility, decisive early action remains rare. The challenge is not technological. It is institutional.

The systems we built to govern risk were designed for a world where danger appeared suddenly and sporadically. In that world, waiting for confirmation was rational. Acting too early could create unnecessary cost, political backlash, or reputational harm. Institutions learned to value restraint. Evidence had to be undeniable before authority could move.

But the nature of risk has changed.

The Institutional Gap

Wildfires, ecological degradation, infrastructure stress, and climate volatility rarely appear as isolated events anymore. They emerge gradually through environmental drift–persistent shifts in conditions that accumulate over time.

Machines can now detect these trajectories with increasing clarity. Models can map patterns that stretch across years, landscapes, and systems. Signals that once felt ambiguous are becoming easier to see.

Yet institutions still struggle to act on them.

What lies between signal and action is legitimacy–the institutional permission to intervene before harm becomes undeniable. In most systems, that legitimacy arrives only after consequences appear.

Modern governance was built to react rather than anticipate. Analysts may recognize converging patterns, but decision structures remain tied to thresholds designed for episodic risk. Action becomes legitimate only when those thresholds are crossed.

Between early signal and formal authority lies a narrow institutional gap where prevention quietly disappears.

No one intends this outcome. The individuals inside these systems are often highly capable and deeply committed to safety, stewardship, and public trust. But institutional incentives shape behaviour in subtle ways.

Acting early concentrates responsibility.

Waiting distributes it.

When something is prevented, success is invisible. When disaster occurs, response becomes visible, measurable, and defensible.

Over time, institutions become extraordinarily good at response while remaining structurally hesitant about prevention.

Artificial Intelligence as Institutional Memory

One of the most important contributions of Artificial Intelligence is not prediction, but memory.

Human attention is episodic. Organizations reset between reporting cycles, leadership transitions, and operational pressures. Patterns that unfold across years often dissolve into background noise.

Machines do not forget.

They retain continuity across time, detecting recurring anomalies and slow structural shifts that humans are forced to normalize. What people experience as vague unease becomes structured evidence in machine perception.

This does not make machines wiser than humans. It makes them differently capable.

Artificial Intelligence extends the temporal horizon of institutional awareness. It allows organizations to perceive trajectories rather than events, convergence rather than isolated anomalies.

But perception alone is not enough.

The real test lies in whether institutions are prepared to trust what they can now see, and whether early signals can be translated into legitimate decisions before crisis provides permission.

The Economics of Prevention

Prevention has always struggled to attract funding because its success is inherently invisible. When prevention works, nothing happens.

No loss appears on a balance sheet. No crisis validates the decision. The outcome exists only as a counterfactual–a story about what did not occur.

But this logic begins to change as volatility intensifies.

When environmental risk becomes persistent rather than episodic, the cost of waiting begins to surface elsewhere. Insurance premiums rise. Capital becomes cautious. Regulatory scrutiny increases. Infrastructure resilience becomes more expensive to maintain.

Nature slowly migrates from the margins of operational awareness into the center of financial decision-making.

At that moment, prevention is no longer simply an environmental or ethical argument. It becomes an economic one.

The institutional question shifts.

Not *"Can we justify acting early?"*

But *"Can we afford not to?"*

Once risk becomes structural rather than episodic, governance must change accordingly. Prevention is no longer a discretionary activity. It becomes a central component of institutional resilience.

Trust: The Invisible Infrastructure

Even with better data and stronger economic arguments, one final barrier remains: trust.

Early action under uncertainty requires a different kind of institutional confidence. Leaders must trust that signals derived from complex systems are meaningful even when visible consequences have not yet appeared.

Trust becomes the invisible infrastructure of anticipatory governance.

Without trust, institutions fall back on the only authority they recognize: events. Action waits until damage provides proof.

But events are the most expensive way to learn.

Building trust requires transparency–clear logic for how signals are interpreted, how decisions are justified, and how accountability is shared. It requires institutions to accept that acting early may occasionally appear unnecessary in hindsight.

Yet waiting for certainty in a world of persistent risk is becoming increasingly costly.

Leadership in the Age of Early Signals

If this book leaves one lasting message, then it is that leadership itself must evolve alongside technology.

Leaders of the past were often defined by how effectively they responded to crisis. They mobilized resources, coordinated institutions, and restored stability after disruption.

Those skills remain essential. But the defining test of leadership in an age of persistent risk will be different.

It will be the ability to act ***before crisis provides permission.***

This requires intellectual courage. Acting early may appear unnecessary in the moment. It may attract criticism from those who see only present conditions rather than emerging trajectories.

Yet the alternative–waiting until consequences become undeniable–is becoming increasingly untenable.

Leadership in the age of early signals therefore requires a shift in mindset:

- From reacting to events to governing trajectories.
- From defending decisions after damage to explaining them before it occurs.
- From seeking certainty to managing convergence.

A Responsibility Shared Across Systems

The responsibility for early action cannot rest with any single actor.

Governments must adapt regulatory frameworks so that acting early is legitimate rather than suspect. Corporations must integrate ecological intelligence into financial and strategic decisions. Investors must recognize resilience not merely as a cost, but as a form of long-term value creation.

Technologists must design systems that make complex signals interpretable and trustworthy.

Artificial Intelligence may provide perception, but perception alone cannot create change. Legitimacy, governance, and capital must move alongside it.

Only when these elements align can early intelligence translate into real-world impact.

Listening to the Landscape

Perhaps the most important lesson of this book comes not from technology but from nature itself.

Ecosystems have always communicated through patterns. Changes in vegetation, water flows, and animal behavior often signal environmental stress long before catastrophe occurs.

Human institutions simply learned to ignore these signals until consequences became unavoidable.

Artificial Intelligence does not replace ecological intelligence. In many ways, it helps us rediscover it. By preserving patterns across time and systems, AI allows us to perceive connections that landscapes have long expressed.

The question now is whether institutions will finally learn to listen.

The Final Responsibility

In the years ahead, early signals will become more visible than ever before. AI systems will continue improving their ability to detect convergence across environmental, operational, and financial domains. The distance between perception and prediction will continue to shrink.

But the most important question will remain human.

Will Leaders Act When the Story Is Still Forming?

The fires of the future may still burn. Risk will never disappear entirely. But the difference between catastrophe and resilience will increasingly depend on whether societies learn to act during the quiet moment before consequence–when intervention is still affordable and prevention is still possible.

That moment rarely looks dramatic.

It looks ordinary.

It looks like a stretch of time where nothing appears to be happening.

And yet it is the moment that matters most.

Because that is where the future quietly decides whether it will become a crisis or remain a possibility that never had to happen.

The responsibility of foresight begins there.

Biography

Dr Dakshinamurthy Venkata Kolluru is a distinguished leader in global digital technology education, with over three decades of experience as a researcher, academician, scientist, mentor, and technology entrepreneur. He began his academic journey at NIT Trichy and went on to earn his master's and Ph.D. from Carnegie Mellon University, USA.

Dr. Murthy has conducted over 200 workshops for CXOs and senior executives across the US, Middle East, India, and Europe, delivering impactful AI-driven solutions that have enhanced business outcomes.

A recipient of the Binani Gold Medal for his contribution to India's first Intercontinental Ballistic Missile under the mentorship of Dr. A.P.J. Abdul Kalam, he has also authored numerous publications, patents, and AI products. He is recognized among the Top 10 Analytics Academicians in India.

Glossary

Understanding the Language of Early Risk and Prevention

The ideas in this book rely on a specific vocabulary–one that reflects how modern risk emerges, how institutions interpret it, and how decisions are made under uncertainty.

This glossary is not a technical dictionary. It is a guide to the concepts that shape how we see, understand, and act before consequence becomes inevitable.

Adaptive Normalisation

The gradual adjustment of systems and people to worsening conditions, where what was once considered abnormal becomes accepted as the new baseline.

Ambiguity Zone

The phase where signals are present but not yet strong enough to justify action. This is where prevention is possible–and most often missed.

Anticipation Window

The limited period during which early intervention is still possible without major disruption. Once closed, only response remains.

Anticipatory Intelligence

The ability of systems–human or machine–to recognise emerging risk before it becomes visible as consequence.

It is not prediction of events, but recognition of patterns that indicate direction.

Authority Threshold

The level of certainty required before institutions feel justified to act. In many systems, this threshold is set too high for modern risk.

Capital Confidence

The willingness of financial stakeholders to support decisions under uncertainty. It depends less on prediction accuracy and more on clarity of decision logic.

Convergence

The point at which multiple independent signals–environmental, operational, financial, and behavioural–begin to align.

Individually, they appear manageable. Together, they remove ambiguity.

Convergence Visibility

The ability to recognise when multiple weak signals are forming a coherent pattern. Often achieved through integrated data and AI systems.

Decision Drag

The slowing of decision-making caused by competing logics across operations, governance, and finance.

Not disagreement–but misalignment.

Decision Legitimisation

The process through which an action becomes acceptable across governance, regulatory, and financial contexts.

Without this, even correct decisions remain unexecuted.

Design Lag

The gap between how systems are designed to operate and how risk actually behaves in the real world.

Distributed Risk

Risk that exists across multiple systems, domains, or geographies, without a single point of ownership.

Early Intervention Cost

The relatively low cost of acting before conditions escalate–often invisible because it prevents measurable events.

Early Signal

A weak but persistent indication that conditions are shifting away from normal behaviour. Rarely urgent in isolation, but meaningful over time.

Ecosystem Intelligence

The natural signals generated by environmental systems such as vegetation stress, animal movement, and moisture patterns that indicate changing conditions.

Endurance Trap

A situation where systems continue to function despite increasing strain, masking the need for early intervention.

Event Bias

The tendency of institutions to prioritise visible events over gradual pattern formation when making decisions.

Explainability

The ability of an AI system to make its reasoning understandable to human decision-makers. Without explainability, intelligence does not translate into trust.

Governance Lag

The delay between when risk becomes visible and when institutions are authorised to act. A structural gap, not a failure of awareness.

Invisible Accumulation

The slow build-up of risk that does not trigger alarms but steadily reduces system resilience.

Legitimacy

The condition under which action is considered justified, defensible, and acceptable across stakeholders. In most systems, legitimacy arrives after consequence when it is already too late.

Legitimacy Gap

The disconnect between what is known through data and what is authorised through governance.

L-ARC™ (Lahiri AI–Resilience Continuum)

A framework that connects early signal detection, interpretation, institutional legitimacy, capital confidence, and action. It reflects the continuous loop required to move from intelligence to impact.

Monitoring Loop

A cycle where signals are continuously observed but not acted upon, often replacing decision- making with surveillance.

Near-Decision

A moment where action is considered but not taken due to insufficient justification or confidence.

Pattern Memory

The ability of systems (especially AI) to retain and recognise long-term patterns that humans may forget or normalise.

Persistence

A defining characteristic of modern risk.

Not intensity, but duration–signals that remain just below thresholds, long enough to reshape outcomes.

Pre-Event Phase

The period before an incident occurs, where signals are present but consequences have not yet materialised.

Prevention Gap

The space between what is known early and what is acted upon. It exists not because of missing data, but because of missing permission.

Prevention Value

The economic and strategic benefit derived from avoiding an event–often difficult to measure because success leaves no trace.

Response Bias

The structural tendency to invest more in responding to crises than in preventing them.

Signal Fatigue

The reduced sensitivity to early warnings caused by repeated exposure to signals that do not immediately result in visible outcomes.